Soft Machines

Nanotechnology and Life

Richard A. L. Jones

Department of Physics and Astronomy
University of Sheffield

OXFORD

UNIVERSITY PRESS

OXFORD

UNIVERSITY PRESS

Great Clarendon Street, Oxford OX2 6DP

Oxford University Press is a department of the University of Oxford.
It furthers the University's objective of excellence in research, scholarship,
and education by publishing worldwide in

Oxford New York

Auckland Cape Town Dar es Salaam Hong Kong Karachi
Kuala Lumpur Madrid Melbourne Mexico City Nairobi
New Delhi Shanghai Taipei Toronto

With offices in

Argentina Austria Brazil Chile Czech Republic France Greece
Guatemala Hungary Italy Japan Poland Portugal Singapore
South Korea Switzerland Thailand Turkey Ukraine Vietnam

Oxford is a registered trade mark of Oxford University Press
in the UK and in certain other countries

Published in the United States
by Oxford University Press Inc., New York

British Library Cataloguing in Publication Data

Data available

Library of Congress Cataloging in Publication Data

Data available

Typeset by Newgen Imaging Systems (P) Ltd., Chennai, India
Printed in Great Britain
on acid-free paper by
Ashford Colour Press Ltd, Gosport, Hampshire

ISBN 978–0–19–852855–5 (Hbk.) 978–0–19–922662–7 (Pbk.)

1 3 5 7 9 10 8 6 4 2

Preface

Nanotechnology, as both a word and a concept, was first popularised by K. Eric Drexler. The power of his concept is proved by the way it has spread beyond the academic and business worlds into popular culture. But as the idea has spread, it has mutated; it now encompasses both incremental developments in materials science and the futuristic visions of Drexler. The interested onlooker could be forgiven some confusion when confronted by the diversity of what is currently being written about the subject.

My aim in writing this book was to re-examine the vision of a nanotechnology that is comprised of tiny, nanoscale machines and engines, but to focus on the question of what the appropriate design rules should be for such a technology. Should we attempt to duplicate, on a smaller scale, the principles that have been so successful for our engineering achievements on a human scale? Or should we try to copy the way biology operates?

To answer this question we need to find out something about the alien world of the nanoscale, where the laws of physics operate in unfamiliar and surprising ways. We need to explore how cell biology operates, and to understand how the design choices that evolution has produced are constrained by the peculiarities of physics at the nanoscale. Then we can begin to appreciate how we should build synthetic systems that achieve some of the same goals as biological nano-machines.

My views on nanotechnology have been developed with the help of many of my professional colleagues. At Sheffield University, my collaborator Tony Ryan has contributed a great deal to the development of the ideas in this book. Amongst my other colleagues at Sheffield, I owe particular thanks to Mark Geoghegan, David Lidzey and Martin Grell. I learnt how important it was for a physicist to try and understand something about biology from Athene Donald and Sam Edwards, at Cambridge. I learnt a great deal about the broader context of Nanoscale Science and Technology from helping to set up a Masters course with that name, run jointly by Leeds and Sheffield Universities. Amongst all those who have been involved with this course I'm particularly indebted to Neville Boden for having the vision to get the course established and to Rob Kelsall for his persistence and attention to detail in managing it. I was given an impetus to think about the wider implications of nanotechnology for society as a result of an invitation by Stephen Wood, of the Institute of Work Psychology at Sheffield, to help write a report on the Social and Economic Challenges of Nanotechnology for the UK's Economic and

Social Research Council, and I'm grateful to him and our co-author Alison Geldart for helping me see what the subject looks like through a non-scientist's eyes. There is, of course, a huge international scientific effort in nanotechnology at the moment. I am very conscious that in picking out individual results and scientists I have omitted to mention a great number of other equally deserving workers round the world. To anyone offended by omissions or mis-attributions of credit, I offer my apologies in advance.

I owe thanks to Dr Sonke Adlung at the Oxford University Press for encouraging me to embark on and persist with this project. Finally, I am indebted to my wife Dulcie for her advice, support, encouragement and much else.

Richard A. L. Jones
Stoney Middleton, Derbyshire
May 2004

Contents

1
Fantastic voyages

A new industrial revolution?

Some people think that nanotechnology will transform the world. Nanotechnology, to these people, is a new technology which is not with us yet, but whose arrival within the next fifty years is absolutely inevitable. Once the technology is mastered, we will learn to make tiny machines that will be able to assemble anything, atom by atom, from any kind of raw material. The consequences, they believe, will be transforming. Material things of any kind will become virtually free, as well as being immeasurably superior in all respects to anything we have available to us now. These tiny machines will be able to repair our bodies from the inside, cell by cell. The threat of disease will be eliminated, and the process of ageing will be only a historical memory. In this world, energy will be clean and abundant and the environment will have been repaired to a pristine state. Space travel will be cheap and easy, and death will be abolished.

Some pessimists see an alternative future—one transformed by nanotechnology, but infinitely for the worse. They predict that we will learn to make these immensely powerful but tiny robots, but that we will not have the wisdom to control them. To the pessimists, nanotechnology will allow us to make new kinds of living, intelligent organisms, who may not wish to continue being our servants. These tiny machines will be able to reproduce, feed, and adapt to their environment, in just the same way as living organisms do. But unlike natural organisms, they will be made from tough, synthetic materials and they will have been carefully designed rather than having emerged from the blind lottery of evolution. Whether unleashed on the world by a malicious act, or developing out of control from the experiments of naïve scientists, these self-replicating nanoscale robots will certainly break out of our custody, and when this happens our doom is assured. The pessimists think that life itself will have no chance in the struggle for supremacy with these nanobots; they will take over the world, consuming its resources and rendering feebler, carbon-based life-forms such as ourselves at best irrelevant, and at worst extinct. In this scenario, we humans will accidentally, and quite possibly with the best of intentions, use the power of science to destroy humanity.

What is now not in dispute is that scientists have an unprecedented ability to observe and control matter on the tiniest scales. Being able to image atoms and molecules is routine, but we can do more than simply observe; we can pick molecules up and move them around. Scientists are also understanding more about the ways in which the properties of matter change when it is structured on these tiny length scales. Technologists are excited by the prospects of exploiting the special properties of nano-structured matter. What these properties promise are materials that are stronger, computers that are faster, and drugs that are more effective than those we have now. Government research funds are flooding into these areas, and start-up companies are attracting venture capital with a vision of nanotechnology that is, perhaps, incremental rather than revolutionary, but which in the eyes of its champions will drive another burst of economic growth in the developed countries. For this kind of enthusiast, usually to be found in government departments and consultancy organisations, nanotechnology is not necessarily going to transform the world; it is just going to make it somewhat more comfortable, and quite a lot richer.

There are some who are simply suspicious of the whole nanotechnology enterprise. They see this as another chapter in a long saga in which different branches of science are hijacked and misused by corporate and state interests. The results will be new products, certainly, but these will be products that no one really needs. The rich will be persuaded by clever marketing to buy expensive cosmetics and ever more sophisticated consumer gadgets, while the poor people of the world continue to live in poverty and ill health. The environment will be further degraded by new nano-materials, even more toxic and persistent than the worst of the chemicals of the previous industrial age.

Some people doubt whether nanotechnology even exists as a single, identifiable technology. We might well wonder what nanotechnology actually is. Is it simply a cynical rebranding of chemistry and materials science, or can we really map out a path from the mundane but potentially lucrative applications of nanoscale science of today to the grand visions of the nanotechnology enthusiasts? Many distinguished scientists are certainly deeply sceptical that the vision of self-replicating nano-robots is achievable even in principle, and they warn that the dream of radical nanotechnology is simply science fiction.

But the visionaries of radical nanotechnology have one unbeatable argument with which to respond to the scepticism of scientists and others. A radical nanotechnology must be possible in principle, because we are here. Biology itself provides a fully-worked-out example of a functioning nanotechnology, with its molecular machines and precise, molecule by molecule, chemical syntheses. What is a bacteria if not a self-replicating, nanoscale robot? Yet the engineering approach that radical nanotechnologists have proposed to make artificial nanoscale robots is very different to the approach taken by life. Where biology is soft, wet, and floppy, the structures that radical nanotechnology envisions are hard and rigid. Are the soft machines that life is built from the unhappy consequence of the contingencies of evolution? When we build a new, synthetic nanotechnology by design, will our creations be able to overcome the frailties of life's designs? Or does life provide us with a model for nanotechnology that we should try and emulate—are life's soft machines simply the most effective way of engineering in the unfamiliar environment of the very small?

This is the central, recurring question of this book. To engage with it, we need to find out in what way the world on the nanoscale is different to the one in which we live our everyday lives, and the extent to which the engineering solutions that evolution has produced in biology are particularly fitted for this very different environment. Then, perhaps, we will be in a position to find our own solutions to the problems of making machines and devices that work at the nanoscale.

The radical vision of nanotechnology

In *Dorian*, Will Self's modern reworking of Oscar Wilde's fable *The picture of Dorian Gray*, the central character is a dissipated hedonist who magically keeps his youthful appearance despite the excesses of his life. At one point, he explores cryonic suspension as a way of staying alive for ever. In a dingy industrial building on the outskirts of Los Angeles, Dorian Gray and his friends look across rows of Dewar flasks, in which the heads and bodies of the dead are kept frozen, waiting for the day when medical science has advanced far enough to cure their ailments. One of Dorian's friends is sceptical, pointing out that the remaining water will swell and burst each cell when it is frozen, and he doubts that technology will ever advance to the point at which the body can be repaired cell by cell.

—'*Course they will, the Ferret yawned; Dorian says they'll do it with nannywhatsit, little robot thingies—isn't that it, Dorian?*
—*Nanotechnology, Fergus—you're quite right; they'll have tiny hyperintelligent robots working in concert to repair our damaged bodies.*

This is the way in which the idea of nanotechnology has entered our general culture. This vision has a single source, K. Eric Drexler's 1986 book *Engines of creation*. Drexler imagined a technology in which factories would be shrunk to the size of cells, and equipped with nanoscale machines. These machines would follow a program stored on a molecular tape, and would be able to build anything by positioning atoms in the right pattern. Drexler calls the machines 'assemblers', and the vision of assembler-based technology 'molecular manufacturing'. Of course, if the assemblers can build anything, then they can build copies of themselves—such machines would be self-replicating.

Drexler's vision of assemblers had two origins. On the one hand, molecular biology and biochemistry shows us astounding examples of sophisticated nano-machines. Consider the ribosome, the machine that synthesises protein molecules according to the specification coded in an organism's DNA—this looks very much like Drexler's picture of an assembler. On the other hand, he drew on a famous lecture given in 1959 by the iconic American physicist Richard Feynman, 'There's plenty of room at the bottom', to stress that there were no fundamental reasons why the trends toward miniaturisation that were driving industries like electronics could not be continued right down toward the level of atoms and molecules. Drexler put these two lines of thought together. What would happen if you could create nano-machines that did the same sorts of things as the machines of biochemistry, but which, instead of using the materials that the chance workings of nature had provided biochemistry, used the strongest and most sophisticated materials that science could provide? Surely you would have a nanotechnology that was as far advanced from the humble workings of a bacteria as a jumbo jet is from a sparrow.

With such a powerful nanotechnology, the possibilities would be endless. Instead of factories building cars and aircraft piece by piece, nanotechnology would make manufacturing more like brewing than conventional engineering. You would simply need to program your assemblers, put them in a vat with some simple feedstocks, and wait for the product to emerge. No matter how intricate the product, with nanotechnology it would be barely more expensive to produce than the cost of the raw materials.

If nano-machines can build things from scratch, then they can also repair them. If you regard the results of disease and ageing as simply being a consequence of misarranged patterns of atoms, then the assembler gives you a universal panacea. Drexler envisaged nano-machines as functioning both as drugs of unparalleled power and as surgeons of unsurpassed delicacy. He did not shrink from the ultimate conclusion—that nanotechnology would allow life to be extended indefinitely. For those who cannot wait for science's slow advance to bring us to this point, there is always the option of putting your body into cold storage and waiting for science to catch up.

What will a future transformed by nanotechnology look like? Many science fiction writers have made an attempt to describe such a future. *The diamond age* by Neal Stevenson is a rich and quite convincing picture, but for all its nuances he presents a world that is more or less a natural extension of modern technological capitalism. Life would be extremely comfortable for the well born and well educated, but considerably less wonderful for those who drew life's less lucky lottery tickets. Meanwhile, there is another, much more terminal view of what nanotechnology might do to us—the dystopian vision of a world taken over by *grey goo*.

It must have been a slow news day in Britain on 27 April 2003, because the lead headline in *The Mail on Sunday*, a mass-market newspaper of rather conservative character, was about science. Characteristically, the story had a royal angle too; the heir to the British throne, Charles, Prince of Wales, was reportedly very worried about the threat posed by nanotechnology. Scientists were risking a global catastrophe in which an unstoppable plague of maverick self-replicating nano-machines consumed the entire world. As an apocalyptic vision, it certainly beat *The Mail on Sunday's* usual fare of collapsing house prices and disappearing pensions, but as a story it was rather older. Drexler's own book, *Engines of creation*, warned of a potential dark side to his otherwise utopian dream. What would happen, if having created intelligent, self-replicating nano-robots, these robots decided that they were not happy with their terms of employment? The result would be the destruction and consumption of all existing forms of life by the nanobots—the world will have been taken over by grey goo.[1]

Although what has come to be known as 'the grey goo problem' was discussed by Drexler, what raised the issue to prominence was the publication, in *Wired* magazine, of an article by Bill Joy, the former chief scientist of Sun Microsystems. At the time, the year 2000, *Wired* was the standard-bearer of West Coast technological triumphalism. The article, however, called 'Why the future doesn't need us', painted a grim picture of a future in which advances in robotics, genetic engineering, and nanotechnology rendered humans at best irrelevant, and at worst extinct. The article is very personal, very thoughtful, very wide ranging, and it carries the conviction of an author who knew at first hand both the rapidity of the progress of technology in recent times and the unpredictability of complex systems.

From the *Wired* article, the dangers of nanotechnology slowly permeated into the public consciousness. The article explicitly linked genetic modification (GM) to nanotechnology as twin technologies with similar risks. So, not unnaturally, those activist groups which had cut their teeth opposing GM started to see nanotechnology as the next natural target. After all, the novelist Michael Crichton, who in the novel *Jurassic Park* had so memorably depicted the downside of our ability to manipulate genetic material, chose nanotechnology as the subject of his novel *Prey*.

What has been the scientists' reaction to the growing fears of grey goo? There has been some fear and anger, I think; many scientists watched the controversy about genetic modification with dismay, as in their eyes a hugely valuable, as well as fascinating, technology was hobbled by inaccurate and irresponsible reporting. But mostly the reaction is blank incomprehension. At least genetic modification was actually a viable technology at the time of the controversy, while for a self-replicating nano-machine there is still a very long way to go from the page of the visionary to the laboratory or factory. To a scientist, struggling maybe to get a single molecule to stick where it is wanted on a surface, the idea of a self-replicating nano-robot is so far-fetched as to be laughable.

How have we got to this state, where we have a backlash to a technology that has not yet arrived? In this, maybe scientists are not entirely without blame. Most scientists working in nanotechnology themselves may refrain from making extreme claims about what the science is going to deliver, but (with some notable exceptions) they have not been very quick to lower expectations. One does not have to be very cynical to link this to the very favourable climate for funding that nanoscale science and technology has been enjoying recently.

---

[1] Why grey? Apart from the appealing alliteration, presumably because the nanobots that make up the goo are made of diamond-like carbon.

I do not think that grey goo represents a serious worry, either, but I do think that it is worth thinking through the reasoning underlying the fears. This is because I believe that this reasoning, deeply flawed as it is, betrays a profound underestimation of the power of life itself and the workings of biology, and a complete misunderstanding of the way that nature works on the nanoscale. Until we clear up these misunderstandings we are not going to be able to harness the power that nanotechnology will give us.

In some of the most extremely optimistic visions of nanotechnology, there is a distrust of the flesh and blood of the biological world that is almost Augustinian in its intensity. This underestimation of biology underlies the thinking that produced the grey goo dystopia too. Surely, the argument goes, as soon as human engineers start to engineer nanobots, the feeble biological versions will not stand a chance. After all, when an evolutionary superior species invades the ecological niche of an inferior one, the inferior one is doomed to extinction. In this cartoon view of Darwinism, dumb dinosaurs were outsmarted by quick-thinking mammals, and hapless Neanderthals were inexorably pushed out by our own *Homo sapiens* ancestors. A similar fate is inevitable when our type of life—basically assembled by chance from all sorts of unsuitable materials patently lacking in robustness—meets something that has been properly designed by a college-trained nanotechnologist. What chance will a primitive bug, little more than a water-filled soap bubble, have when it meets a gleaming diamond nanobot with its molecular gears grinding and its nanotube jaws gnashing?

Of course, there are no primitive bugs (at least on Earth), and we ought to know very well that while individual organisms can seem frail, life itself is spectacularly tough. The insights of molecular cell biology show us more and more clearly how optimised nature's machines are for operation at the nanoscale. If the mechanisms nature uses seem odd and counter-intuitive to us, it is because the physical constraints on design are very different on the nanoscale from the constraints in the world we design for. The other insight that we should take from biology is that evolution is an extremely efficient design principle that works as well—possibly better—at the level of molecules as it does with finches and snails. The biological macromolecules that form the basis of the nano-machines in even the simplest-looking cells have themselves evolved to the point at which they are extremely effective at their jobs.

Surely though, a steam engine is better than a horse, strong and lightweight aluminium alloy is a better material to make a wing out of than feather and bone . . . if we can find materials that are so much better than the ones nature has given us to work at the macroscopic level, then surely the same is true at the nanoscopic level? This is Drexler's argument, but I disagree. Nature has evolved to get nanotechnology right. Most of nature exists at the nano-level, the necessary mechanisms and materials were evolved very early on, work extremely well, and look pretty similar in all kinds of different organisms. Big organisms like us consist of mechanisms and materials that have been developed and optimised for the nanoworld, that evolution has had to do the best it can with to make work in the macroworld. We are soft and wet, because soft and wet works perfectly for bacteria. Because we have evolved from bacteria-like organisms we have had to start with the same nano-machinery and try and build something human-sized out of it. No wonder it seems a bit clunky and inadequate on a human scale. But at the nano-level, it is just right.

Nano everywhere

It is difficult to visit a university anywhere in the world nowadays without falling over a building site where a new institute of nanotechnology or nanoscience is due to open. Taking the lead from the USA, where in 1999 President Clinton announced a National Nanotechnology Initiative, governments and science-funding bodies across the world have been pouring hundreds of millions of dollars into the areas of nanoscience and nanotechnology. Scientists have risen to the challenge, and nanotechnology now forms one of the most active areas of scientific endeavour.

So does this mean that Drexler's vision of molecular manufacturing and nanoscale assemblers will soon be with us? No. It is fair to say that most scientists working in the area of nanoscience and technology regard the Drexlerian program as being somewhere along the continuum between the impractical and the completely misguided. Instead, what we see is a great flowering of chemistry, physics, materials science, and electronic engineering, a range of research programs which sometimes have little in common with each other besides the fact that their operations take place on the nanometre scale. Some of this work, that is now called nanoscience or nanotechnology, is actually no different in character to what has been studied in fields like metallurgy, materials science, and colloid science for the last fifty years. Control of the structure of matter on the nanoscale can often bring big benefits in terms of improvements in properties, and this is the basis of many of the improvements which we have seen in the properties of materials in recent years. One could call this branch of nanotechnology *incremental nanotechnology*.

Perhaps more novel are those areas of science where advances in miniaturisation are being scaled down further into the nanoscale. One might call this area *evolutionary nanotechnology*, and the type example is micro-electronics. Driven by the huge size of the worldwide electronics and computing industries, the technologies for making integrated circuits have matured to the point that feature sizes of less than 100 nm are now routine. Related technologies are used to make tiny mechanical devices—micro-electro-mechanical systems—which already find use in applications like acceleration sensors for airbags. At the moment devices that are in production are characterised by length scales of tens or hundreds of microns rather than nanometres, but very much smaller devices are being made in the laboratory. Other types of evolutionary nanotechnology include molecular electronics—the creation of electronic circuits using single molecules as building blocks, as well as concepts that are being developed for packaging molecules and releasing them on a trigger. These are beginning to find applications for delivering drugs efficiently. In evolutionary nanotechnology we are moving away from simply making nanostructured materials, toward making tiny devices and gadgets that actually do something interesting.

So where does this leave nanotechnology in the radical sense that Drexler suggested? A very small proportion of the scientists and technologists who would claim to be nanotechnologists are working directly toward this goal, and indeed many of the most influential of these nanotechnologists are deeply sceptical of the Drexler vision. Does this mean, then, that *radical nanotechnology* will never be developed? My own view is that radical nanotechnology will be developed, but not necessarily along the path proposed by Drexler. I accept the force of the argument that biology gives us a proof in principle that a radical nanotechnology, in which machines of molecular scale manipulate matter and energy with great precision, can exist. But this argument also shows that there may be more than one way of reaching the goal of radical nanotechnology, and that the path proposed by Drexler may not be the best one to follow.

Into the nanoworld

Nanotechnology gets its name from the prefix of a unit of length, the nanometre (abbreviated as nm), and in its broadest definition it refers to any branch of technology that results from our ability to control and manipulate matter on length scales between a nanometre and 100 nanometres or so. One nanometre is one-thousandth of a micrometre or micron. This in turn is one-thousandth of a millimetre. How can we put these rather frighteningly small numbers into context?

Everyone is familiar with the macroworld, the world of our everyday experience. We can directly touch and interact with objects with sizes from around a millimetre up to a metre. This is our human world, and in it we have an intuitive understanding of how things move and behave.

The microworld is less familiar, but not completely foreign. The tiniest mites and insects have sizes of a few hundred microns (or a few tenths of a millimetre); these are visible to those of us with good eyesight as little specks or motes, but we need a magnifying glass or low-power microscope to see very much of the individuality of these objects. These are the smallest things that we have direct experience of—the thickness of a human hair, the thickness of a leaf of paper; these all represent lengths at the upper end of the microworld, around 100 microns.

The microworld is familiar territory to engineers. Precision measuring instruments, like micrometers and vernier callipers, can easily measure dimensions to an accuracy of tens of microns. The experienced workshop technicians in my university's machine shop still, despite metrication, think in terms of one-thousandth of an inch, 25 microns, as a precision to which they can, without trying very hard, build components for scientific instruments.

Biologists, too, work naturally in the microworld; it is the world that can clearly be seen through a light microscope. The largest single cells, an amoeba, or a human egg cell, are just about visible as specks to the naked eye, around 100 microns in size. But most animal and plant cells fall into the range of sizes between 10 microns and 100 microns. The simplest forms of single-celled life—bacteria—are a little bit smaller. These ubiquitous organisms, a few of which are feared as the agents of disease, are usually around a micron in size. Most bacteria are clearly visible in a light microscope, but they are too small to see very much internal structure within them.

The internal structure of cells belongs to the nanoworld. At these sizes, things are too small to see with a light microscope. But new techniques have, in the last fifty years, revealed that within what a hundred years ago was thought of as an unstructured, jelly-like protoplasm, there is a fantastically complex world of tiny structures and machines. Inside each of the cells in our bodies are structures such as mitochondria, tiny bodies made from convoluted foldings of membranes, like crumpled balls of paper. Inside plant cells are chloroplasts, the structures in which light is collected and turned into useful energy. Smaller still we would see ribosomes, the factories in which protein molecules are made according to the specifications of the genetic code that is stored in DNA.

Now we are down to the level of molecules, albeit rather big ones. Biological nanostructures, such as ribosomes, are made up of very big molecules, such as proteins and DNA itself, each of which is made up of hundreds, thousands, or tens of thousands of individual atoms. A typical protein molecule might be somewhere between 3 and 10 nm in size, and will usually look like a compact but knobbly ball. We can make big molecules synthetically too. Long, chain-like molecules consisting of many atoms linked together in a line are called polymers, and they are familiar to us as plastics. Materials like nylon, polythene, and polystyrene are made up of such long molecules. If we could see a single molecule in a piece of polyethylene, then it would look like a fuzzy ball about 10 nm big. Unlike the protein molecule, this is not a compact lump, it would be more like a loosely-folded piece of string.

Small molecules are made up of a few atoms; from the three that make a water molecule, to the tens of atoms that make up a molecule of soap or sugar. An individual atom is a fraction of a nanometre in size, so these small molecules will be around one nanometre big. It is the size of these small molecules that defines the lower end of the nanoworld.

As we have seen, the nanoworld is now the realm of cell biology, and it is our efforts to make structures and devices on this scale that defines nanotechnology. How far have we come toward achieving this goal?

The technology that has come the furthest by shrinking the most has been the electronics industry. The original electronic computers were very much artefacts of the macroworld. Older readers will remember that, in the 1960s and before, the crucial components of a radio were thermionic valves, devices the size of small light bulbs. Before the introduction of transistors, these were at the heart of both amplifiers and logic circuits. So the first computers consisted of rooms full of racks of electronics, the basic unit of which was the centimetre-sized valve.

It was the invention, firstly of the transistor, but most crucially of the integrated circuit, that allowed electronics to move from the macroworld into the microworld. The transistor meant that electronic components could be made entirely in the solid state, doing away with the vacuum-filled glass bulbs of thermionic valves. The integrated circuit allows us to pack many different electronic components onto a single piece of semiconductor, to produce a complete electronic device in one package—the silicon chip.

In the integrated circuit, single components are not individually hewn from the semiconductors they are made from. Instead, lines etched on the surface of the chip define the transistors that are wired up to make the circuits. How small one can make the components is limited by how fine one can draw lines, and it is a reduction in this minimum line size that has driven the colossal increase in available computer power that we are all familiar with. The minimum line size commercially achievable fell below one micron in the mid 1980s, and is currently well below 100 nm.

We live in the macroworld, we have mature technologies that operate in the microworld, and we are beginning our discovery of the nanoworld. Are there any worlds on even smaller scales that remain to be exploited? There is an old rhyme which captures this sense of worlds within worlds and structures on ever smaller scales: 'Big fleas have little fleas, upon their backs to bite them. And little fleas have littler ones, and so ad infinitum.' But how small can you go? Is there another world that is even smaller than the nanoworld? Physics tells us that there is such a world, the world of subatomic structure. Can we look forward even further to yet more powerful technologies, which manipulate matter on even finer scales, the worlds of picotechnology and femtotechnology?

We now know that atoms themselves, far from being the indivisible objects imagined by the Greek originators of the concept, have a substantial degree of internal structure. Take, for example, a carbon atom. To a chemist, this is an indivisible ball with a diameter of 0.14 nm. It was the achievement of nuclear physics in the early part of the twentieth century to show that the atom was not an indivisible entity; it has internal structure. Ernest Rutherford, a physicist from New Zealand, was able to show in experiments carried out in Manchester that most of the mass of an atom is concentrated at its centre, in a tiny, dense object called the nucleus. This is small, very small—the nucleus of a carbon atom is about 3 femtometres in diameter (a femtometre being *one-millionth* of a nanometre).

But the nucleus is not where the story stops; it itself is made up of protons and neutrons, which themselves have some finite size. The proton can exist independently; since the nucleus of a hydrogen atom consists of a single hydrogen ion—a hydrogen atom with its accompanying electron stripped off—is a free-living proton. Neutrons, too, can exist independently, but not indefinitely; after their lifetime of about ten minutes a free-living neutron will decay into a proton, and an electron and an antineutrino.

For a while it was thought that protons and neutrons were truly fundamental particles, but it turns out that they, too, are composites. Experiments in the 1960s showed that, in exactly the same way as an atom is mostly empty space, with its mass concentrated in a tiny nucleus, protons and neutrons are composed of much smaller particles. Protons and neutrons are each made up of three particles called quarks.

Is there further internal structure to be found, at still smaller lengths, within the quark? It is currently believed that there is not; quarks, and electrons, are believed to be fundamental particles that are not further divisible. Inasmuch as it makes any sense to talk about the size of these particles at all, they have no finite size—they are true points, without extension in space. In passing, it is worth noting that one might object to this proposition, noting that the suggestion that particles exist that are true points causes all sorts of philosophical and physical problems. This is indeed the case, and it is the business of quantum field theory to sort these problems out.

We can manipulate matter on scales below the atomic; this is the business of nuclear physics. This is now a relatively old technology, and one that has been, to say the least, a mixed blessing for humanity. It is possible to rearrange the protons and neutrons within the nucleus to obtain new elements, even elements that are unknown in nature. But the characteristic of these transformations, as well as the transformations of nuclear fusion and nuclear fission, is that they involve very high energies.

Energy scales

There is a relationship between the length scale at which one is trying to manipulate matter and the relative size of the energy input that is needed to make transformations of matter on that length scale. Roughly speaking, the smaller the length scale on which one operates, the higher the energies that are involved in these transformations. This is why, to look inside the very smallest subatomic particles, particle physicists need to build huge accelerators many miles in diameter. Nuclear physicists probe and manipulate the interior of atomic nuclei; their smaller accelerators can be fitted into tall buildings. Chemists, on the other hand, rearrange the peripheral electrons on atoms, and this they can do simply with a Bunsen burner.

The brute force way of putting energy into something is to heat it up, and the temperature of a material is a measure of the amount of available thermal energy per molecule. The transmutations that chemistry can achieve involve the absorption and release of amounts of energy that correspond to temperatures of hundreds or, at the most, thousands of degrees.

But chemical transformations—even the most highly-energetic ones, such as the detonation of explosives—only tinker with the structure of the outermost edges of the structure of atoms. Only the outermost, most loosely-attached electrons are affected by these changes. To rearrange the nucleus, very much higher temperatures are required. If a gas of the heavy isotope of hydrogen, deuterium, can be heated up to a few hundred million degrees in temperature, then pairs of deuterium nuclei can combine to create helium. In the process, they release a great deal of energy. For this reason, nuclear fusion, if it could be controlled, would be able to provide all of our energy needs. The problem is those enormously high temperatures, which are so much greater than any solid material can sustain.

But the temperatures at which nuclear transformations take place—the temperatures at the centres of the sun and stars—are still tiny compared to the temperatures one needs to transform the deep components of the protons and neutrons—the quarks. At a temperature of around 10^{12} K (around 100 000 times hotter than the temperature at the centre of the sun), the quarks that make up protons come apart from each other to make an undifferentiated soup—the quark–gluon plasma. These conditions are thought to have existed very early in the universe, shortly after the big bang.

These conditions involve unimaginably high levels of energy. What of nanotechnology—what are the natural energy scales that characterise transformations that take place within the molecular machines and structures of our own cells? The energies have to be adapted to the temperatures at which we live, a room temperature of 300 K. These are the energies, not of the violent fusions and sunderings of nuclear physics, but of the rather gentle stickiness of a post-it note. Biology is low-energy physics.

Different physics at the nanoscale

It is an axiom of science that the fundamental laws of physics are constant and unchanging; we believe them to be the same for all objects at all times and in all places. But in the working lives of most physicists, and all engineers and technologists, what one is using to predict and control the behaviour of material things are not the fundamental laws of physics, but a set of approximations and rules of thumb that happen to operate in one particular domain. If we are architects designing a building made of stone, then we use the classical laws of statics. These are a subset of the laws of classical mechanics, which we can think of as an approximation to what we believe to be the ultimate laws governing the behaviour of matter, quantum mechanics, which is appropriate for macroscopic objects. Together with these laws, we use some rules of thumb—that stone is incompressible, but that it will break under tension, for example—that we know are not strictly correct, but which are close enough to being right that they allow us to build buildings that do not collapse. What mixture of approximate laws and rules of thumb we should use will be very different on the nanoscale than the ones we are familiar with from the macroworld.

One key difference is the importance of quantum mechanics. In fact, it is becoming a received truth that the difference between the macroworld and the nanoworld is that, while the macroworld is governed by the classical mechanics of Newton, the nanoworld is governed by the mysterious and counter-intuitive laws of quantum mechanics. Like much received wisdom, there is a kernel of truth in this, surrounded by much that is misleading. The real situation is much more complicated than that. To start with, some very familiar, everyday properties in the macroworld can only be properly understood in terms of quantum mechanics. Why metals conduct electricity, why magnets attract iron, why leaves are green . . . classical mechanics provides no explanations at all for these questions, which can only really be understood in terms of quantum mechanics. On the other hand, quite a lot of what is special about the nanoworld does not depend on quantum mechanics. This is particularly true when there is water around, and the temperature is closer to the comfortable warmth of everyday life than the chilly environs of absolute zero that physicists often like to do experiments in.

The big difference between the macroworld and the nanoworld, if we are not at an ultra-low temperature and in a vacuum, but in a water-filled beaker at room temperature, arises from the fact that water (and everything else) is made of molecules. These molecules are constantly flying around at high speed in random directions, hitting whatever happens to be in their way. This leads to a distinctive feature of the nanoworld—Brownian motion. Everything is continually being shaken up and jiggled around.

The other unfamiliar feature of the nanoworld is its stickiness—when surfaces get close, they almost always like to stick to each other. It is inevitable that when you make things smaller their surfaces get more important, so working around this stickiness problem is a central part of the technology of finely-divided matter. This is well known in those traditional branches of science and technology that deal with finely-divided matter. People who make paint devote a lot of attention to making sure that the tiny paint particles stay in suspension and do not form a sticky goo at the bottom of the tin. But the importance of the problem is maybe not fully appreciated by those who sketch designs for nanoscale machines.

It is these unfamiliar features of the nanoworld that make engineering in this domain so unfamiliar and non-intuitive. Imagine mending your bicycle in the shed one day, as a simple example of the kind of everyday engineering we are familiar with. The parts are rigid, and if we screw them in place they stay where we put them. Mending a nano-bicycle would be very different. The parts would be floppy, and constantly flexing and jiggling about. Whenever different parts touched there would be a high chance that they would stick to each other. Also, the pile of screws that we had left in a pot would have jumped out by themselves and would be zigzagging their way toward the garage door. Nanoscale engineering is going to be very different from human-scale engineering, but if we need lessons then we know where to look. The more we learn about the nanoscale mechanisms that biology uses at the level of the cell, the more we learn how well adapted they are for this unfamiliar world. This book begins to look for some of these biological lessons for nanotechnologists.

2
Looking at the nanoworld

The nanoworld was not invented by Richard Feynman or K. Eric Drexler. Long before the idea of nanotechnology was devised and the word coined, technologies and processes that humans depended on relied on the manipulation of matter at the nanoscale, even if the way these technologies produced their effects were not fully understood at the time. Take the invention of Indian ink by the ancient Egyptians or the discovery of how to make soap; both of these long-established materials undoubtedly rely on nanotechnology in the broad sense, and if these inventions were being made today then their inventors would no doubt be stressing their nanotechnological credentials as they attempted to raise capital for their start-up companies. What makes it possible to think of the nanoworld as a new realm of matter that we can explore and control is the availability of instruments that allow us to see into that realm. It only became possible to appreciate the vast extent of the universe beyond the Earth after the telescope had been discovered. So it is, that the invention of new microscopes, capable of picking out the details of the world on scales smaller than a micron, has enabled us to appreciate the scope of the nanoworld. Seeing is believing.

If you want to look at something small, then you need a microscope. When we talk of microscopes and telescopes, this suggests ways of enhancing our sense of vision. This is perhaps natural given how most of us depend on our sense of sight in our interaction with the ordinary world, in the realm of our own senses. Ordinary light microscopes and telescopes are essentially enhancements of our own eyes, whose technology is in direct descent from the medieval invention of spectacles.

The development of the optical microscope in the seventeenth century opened up a new world whose existence had not been previously suspected—a world filled with tiny animals and plants of strange designs, and ultimately of microbes. Many of these microbes were revealed to be beneficial to humanity, like the yeasts that convert grape juice to wine and the bacteria that convert milk to yoghurt. Others are harmful or even fatal, like the pathogens that cause smallpox and the plague. But the majority simply make their living in their own world without much impact on humans. This world that the light microscope reveals we can call the microworld—the world defined by dimensions between a micron or so—the size of the smallest object that a light microscope can discern—and the fraction of a millimetre that can be made out by the unaided naked eye.

That there is another world even smaller than the microworld—the nanoworld—was clear even before the tools required to image it became available. The lower size limit on the nanoworld is set by the size of molecules, and long before molecules could be directly imaged there were indirect ways of estimating their size. At the end of the nineteenth century and the beginning of the twentieth it was becoming apparent that a whole class of matter was made up of objects that were bigger than molecules but still well below a micron in size. Glues and gums, milk and blood, it was clear that these were not just simple solutions like a solution of sugar or salt. The evidence was unequivocal that colloids, as these materials were called, consisted of a dispersion in water of objects that were characterised by nanoscale dimensions. It was not clear at the time whether these nanoscale components were aggregates of smaller molecules or very large individual molecules—macromolecules.

But even the best light microscopes do not let us see into the nanoworld proper. Fundamental physical limits that arise from the wave nature of light mean that it will always be impossible to discern objects with dimensions much less than a micron with a light microscope of conventional design. To extend our vision into the nanoworld, it is necessary to use different kinds of radiation.

The size and shape of molecules were being determined by X-ray diffraction in the first half of the twentieth century. In particular, the existence of very large molecules—macromolecules—was confirmed. X-ray diffraction is a method that could determine the size and structure of molecules directly; after the development of the technique at the beginning of the twentieth century by Max von Laue and the Braggs (Laurence and William, the most famous father and son team in science), the technique was applied to bigger and bigger molecules. By 1950, the significance of macromolecules in biology was clear, and the importance of determining the structure of biological macromolecules was obvious. At this time diffraction patterns had already been obtained for proteins, and most famously, in 1953, the structure of the macromolecule DNA was solved by Francis Crick, James Watson, Maurice Wilkins, and Rosalind Franklin. X-ray diffraction unambiguously tells us not only the overall size of molecules, but also their internal structure . . . but for many scientists the complicated mathematical relationships that relate the diffraction patterns you see on the photographic plate and the structure of the molecules themselves makes the technique less satisfying than being able to visualise something directly with microscopy. More seriously, the technique does rely on being able to make a crystal—a regular three-dimensional repeating array—from the molecule.

So, despite having fairly convincing evidence of the existence of the nanoworld, and something of its richness and complexity, without a better microscope than the optical instruments available in the first half of the twentieth century there was a lack of immediacy about people's knowledge of the nanoworld. Seeing is believing, and even for the most rational scientists there is something much less satisfying about knowledge that is inferred than knowledge obtained from direct observation. So why can we not make microscopes that operate at higher magnifications, to see the nanoworld directly?

Such microscopes can be made, but not using light—you need to use electrons. The electron microscope was invented in 1931, and soon it was illuminating the richness of the nanoworld. Even within the cells of plants and animals there was another whole world of complexity; wheels within wheels in the shape of a whole cast of tiny organelles. Even in such a humble object as a plastic bag there are hierarchies of structures, built on sheaves of macromolecules. And yet, despite these advances, electron microscopy is still a long way away from the immediacy of the light microscope. The instruments are complicated to operate, temperamental even; they are expensive, and the images are not always easy to interpret without years of experience. Delicate samples can be damaged or even obliterated by the huge doses of radiation that the samples are subjected to. Perhaps most importantly, elaborate and complicated procedures need to be gone through to prepare the sample for examination in the microscope. Soft samples need to be mounted, frozen or dried, cut into tiny slices, coated with metals, and then put into the hostile environment of an ultra-high vacuum. It is a long way away from the convenience of being able to put a drop on a slide and peer down a light microscope at it. The electron microscope does at least have the end product that is a photograph, a simple magnified image, rather than the abstract pattern of spots on a photographic plate that X-ray diffraction produces. But the process of preparing the sample and making the image is less like taking a quick look at an object, and more like commissioning an artist to make a painting. There is no question of capturing any motion; everything in the sample has to be commanded to stay still, and you have to live with the knowledge that the likeness of your image is mediated by the quirks of the artist.

It was the invention of an entirely new type of microscope that ultimately led to what we might call the democratisation of the nanoworld—the development of a microscope capable of visualising individual molecules, but without the need for difficult and potentially destructive sample preparation that electron microscopy imposes, and available at a low enough cost that most nanoscience laboratories could afford one, just as they would have a robust light microscope. These new instruments were the scanning tunnelling microscope and the scanning force microscope (or atomic force microscope), both invented within a few years of each other in IBM's Zurich laboratory.

Scanning probe microscopes rely on an entirely different principle to both light microscopes and electron microscopes, or indeed our own eyes. Rather than detecting waves that have been scattered from the object we are looking at, one feels the surface of that object with a physical probe. This probe is moved across the surface with high precision. As it tracks the contours of the surface, it is moved up or down in a way that is controlled by some interaction between the tip of the probe and the surface. This interaction could be the flow of electrical current, in the case of a scanning tunnelling microscope, or simply the force between the tip and the surface in the case of an atomic force microscope. The height of the tip above the sample surface is recorded as it is scanned across the surface, allowing a three-dimensional picture of that surface to be built up on the controlling computer. In going from a light or electron microscope to a scanning probe microscope we have moved away from looking to touching.

Light microscopy

By medieval times, ageing monks were able to use magnifying glasses to help them read the small writing of their manuscripts, and it was almost certainly a spectacle maker (most probably the Dutchmen Hans and Zaccharias Janssen in the late sixteenth century) who realised that two or more lenses could be combined to make a microscope of considerable magnifying power. The way a microscope works is that a lens produces an inverted, real image, which is then magnified by an eyepiece—which works in essentially the same way as a magnifying glass. The power of a microscope lies in the fact that it has more than one stage of magnification. A good high-powered objective might have a magnifying power of 100, while the eyepiece might have a magnifying power of 10. In producing the final image, these magnifying powers are multiplied together; in this case an object one micron big appears to the observer to have the easily discernible size of a millimetre. Can we push the magnifications up yet further so that we can see nanometre-scale objects? There is no reason why we should not introduce another stage of magnification still.

There is a lot of subtlety in the classical optics that describes how a microscope works. But at its simplest, we can think of a microscope as a device that takes the light emitted from a single point on the sample, and maps it onto another single point on a detector, whether that is the retina of a human eye, a piece of photographic film, or (most commonly, nowadays) a charge-coupled device—an electronic detector of the sort that you have in a digital camera or video camera. In the ray optics that you use to analyse a microscope, you draw lines representing the rays of light as they travel from the sample to the detector, being bent by lenses and blocked by apertures as they make their journey. If the microscope has been designed correctly, then no matter which path the light takes from the sample to the detector, light that leaves one point on the sample arrives at one point on the detector; light that leaves two adjacent points on the sample, separated by some small distance, will arrive at adjacent points on the detector, separated by a larger distance which is simply the distance separating the points on the sample multiplied by the magnification of the microscope. In a less-well-designed microscope, or one made with components such as lenses of lower quality, light leaving a single point on the sample arrives not on one single point on the detector, but on a little area. If the areas arising from light coming from two closely-spaced features on the sample overlap, then we will not be able to distinguish those features. Now our image is not perfect, but blurred. Obviously, the aim of microscope design must be to reduce this blurring effect to the minimum possible.

The limit to this process of design is that the basic picture that it is based on, of light travelling in rays from one point to another, is incorrect, or at least incomplete. We have known for a couple of centuries that light is a wave, not a ray, and because of this even a beam which has been most carefully prepared to be parallel will slowly spread out. Lasers, for example, can emit a beam of light that can be made very parallel, so the spot the beam makes if it hits a screen stays very small, even if the screen is a long way from the laser. But spread out the beam certainly does, no matter how carefully we make the beam parallel (collimate it, to use the technical term). As we get further and further away from the laser, the edges of the spot get less and less sharp, and eventually the beam broadens and becomes fuzzier. The cause of this effect is diffraction, and its effect is that, even in the best-designed microscopes, there is a fundamental lower limit on the size of features that can be distinguished, or resolved. This lower limit is related to the wavelength of light; as this varies from about 400 nm (for violet light) to 700 nm (for red light) this means that it is fundamentally impossible to use a light microscope to image the nanoworld.

Before considering how more sophisticated microscopes can look down further into the nanoworld, it is worth remarking that a simple light microscope is still a remarkably powerful and useful tool for the nanoscientist. There are some important features of the nanoworld that are discernible at the length scales of a light microscope—I am thinking here particularly of the phenomenon of Brownian motion, which we will discuss in detail in Chapter 4, and which plays such an important role in understanding why the nanoworld is so different to the macroscopic world. This is easily observable at home with a hobbyist's microscope, or with the sort of microscope available in schools. This illustrates one of the advantages of light microscopes over other, apparently more powerful techniques with higher available magnifications and resolutions. A light microscope can observe a system as it is, a living system for example, without any special treatment, and you can see how things move around as well as seeing their frozen structure.

Perhaps the most important advantage of using light is that there is a lot of it about, and it is easy to detect. It is not difficult to obtain light from very high intensity sources—discharge lamps, lasers—and it is possible to detect light at very low levels. Even our own naked eyes are very efficient detectors of very dim light (and mammals like cats, adapted for life at night, are much better at this than we are), and modern semiconductor-based detectors can detect a single unit of light—a single photon. Taken together, the ease of generating very intense beams of light, and the efficiency with which we can detect very dim signals, means that if we are trying to visualise something very small, then we should not be limited by the fact that this small object only reflects a very small amount of the light that is incident on it. As we shall see next, this means that it is possible to use a light microscope to see a single molecule, even though the wavelength-limited resolution of a light microscope means that we do not see it at its true size.

Seeing a single (big) molecule

I do not know who the first person to see a single molecule using a light microscope was, but I remember very clearly the single-molecule experiment that first made an impression on me. The subject it addressed was the question of how a polymer molecule could move around in a melt of other polymer molecules. Think of a pan full of spaghetti, full of long, flexible, slippery strands, and ask how can one of these strands, tangled up as it is with all the other strands, move through the pan? The problem is an important one for understanding how molten polymers—like the molten polyethylene that is extruded or moulded to form plastic bags or washing-up bowls—can flow at all. The theoretical physicists Pierre Gilles de Gennes, from Paris, and Sam Edwards, from Cambridge, had the insight that the only way a single polymer molecule could move through such a tangle was if it wiggled head first, its body following the path its head made, like a snake moving through long grass. De Gennes coined the term reptation to describe this kind of motion, and, on the basis of this rather pictorial insight, de Gennes, Sam Edwards, and the Japanese theorist Masao Doi developed a complete and quantitative theory of polymer motion. But how could the theory be proved? There was a lot of indirect evidence—the theory seemed to correctly explain the way the flow characteristics of polymers depended on how long the chains were, for example—but the direct proof, that would convince the sceptics, was still missing. Then, on the cover of *Science* magazine, was a series of images that showed a long molecule starting out in the shape of a letter R, and then moving in exactly the snake-like way that de Gennes and Edwards had imagined, its tail following the sinuous path that its head had made through the (invisible) surrounding forest of other molecules.

The images came from a paper by Stephen Chu, a Stanford physicist who, soon after this paper, won the Nobel prize for physics. But what was perhaps most remarkable about them was that they had been made, not with some fabulously expensive and sophisticated piece of new scientific equipment, but with a plain old light microscope. What were the secrets that allowed this remarkable feat?

The first point is that Chu made life easy for himself by finding a big molecule to look at. He chose DNA, the polymer that carries the genetic code. We will say a lot more about this molecule later, but for now what is important is that it is very long, and very stiff. DNA molecules can easily be tens or even hundreds of microns long—an extraordinarily large figure for a single molecule—and because the molecule is so stiff it does not coil up on itself. Instead, it tends to form grand, sinuous curves. This means that, even though Chu's microscope was still bound by the diffraction limit, the molecule was long enough to be resolved. Of course, although the molecule is very long, it is still only a few nanometres wide, and the light microscope cannot resolve this dimension. So, instead of seeing a very fine, long line, what the microscope seems to show is a rather fuzzy sausage shape.

The second thing that Chu did to make the experiment possible was that, rather than using light scattered or reflected from the molecule to make the image, he used its fluorescence. Fluorescence is an optical process in which some molecules, if they are illuminated by light of one colour, re-emit the light with a different colour. If you have ever worn a white shirt at a party with ultraviolet lighting then you have seen the effects of fluorescence—the ultraviolet light is invisible to the human eye, but fluorescent dyes that are incorporated in washing powder ('for brighter than bright whites') make your shirt glow with visible light. The advantage of fluorescence for the microscopist is that one can illuminate the sample with a very bright light source—for example, a laser—and then use a coloured filter to block any of this light from being scattered into the eyepiece. Since the light that the fluorescent molecule is emitting is a different colour to the light used to illuminate the sample, it passes through the filter. In this way, because the background is dark, even the very weak signal from a single molecule can still be picked out.

DNA by itself does not naturally fluoresce. Before one can use this technique one needs to attach fluorescent dye molecules to the DNA—to label it, in effect (see figure 2.1). This could potentially cause difficulties—one needs to find a dye which will stick to the DNA, and one can ask valid questions about whether the behaviour of the molecule might be affected by having these

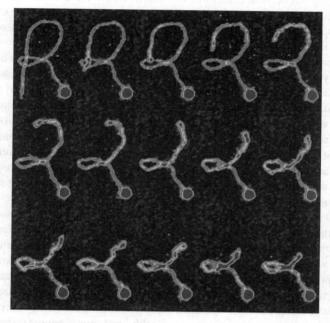

Fig. 2.1 Pulling a single DNA molecule. One DNA molecule has been fluorescently labelled, and attached to a micron-sized polystyrene bead. The bead is moved using 'laser tweezers' through a solution of non-labelled DNA. The images are obtained from optical fluorescence microscopy. See T. T. Perkins, D. E. Smith, and S. Chu. *Science* **264** (1994) 819; this figure by courtesy of Steve Chu.

dye molecules stuck to it. But the need to label can be used to advantage, too, and this is what Chu's experiment relied on. Imagine once again our pan of spaghetti, and suppose that we are looking at the pan with vision that is too blurred to see the width of each piece of spaghetti sharply. If all the pieces of spaghetti are glowing, then all we will see is an undifferentiated block of colour. But if we have taken out one single strand of spaghetti and labelled it with luminous paint before returning it to the pan, then if we look at the pan we will clearly see our labelled strand, glowing against the dark background of the unlabelled strands. This is what Chu did in his experiment; a very low concentration of DNA was fluorescently labelled, leaving most of the DNA in his sample unlabelled, and thus invisible.

If you want to look at single molecules with light microscopy, they need to be highly dilute, so the blurred signals from each molecule do not overlap with each other. So an experiment like Chu's, whose point is to study the way the molecules interact with each other, is only possible by using selective labelling.

Fluorescence microscopy is an important tool in biology, because it allows one to begin to see something of the complicated traffic of molecules within the cell, as well as the static structure of the cell itself. The power of using fluorescent labelling has become even more obvious thanks to the introduction of a new kind of light microscope—the scanning laser confocal microscope. Confocal microscopes suffer from the same fundamental limitations on resolution that an ordinary light microscope has, but have some powerful advantages.

Anyone who has played with a microscope knows the importance of focusing it. If you are looking at a surface, you rotate the focus knob, which physically moves the microscope away from the surface until the blurred features snap into clarity. If you are looking, not at a surface, but at a transparent sample, what you see is just one section of the sample, the thin slice that happens to be at the right distance to be in focus. But the light that comes from the out-of-focus parts of the sample still enters the eyepiece; if you were doing fluorescence microscopy this would give a diffuse glow that would reduce the contrast with which the in-focus parts of the image would stand out from the background.

What a confocal microscope does is remove the glow from the out-of-focus parts of the sample. To achieve this, it has to work in a completely different way to a normal microscope. Instead of illuminating a wide area of the sample and forming an image of this whole area, a laser beam is finely focused down to an intense point, and as this little point of light is scanned across the sample the detector signal is recorded to build up an image. In this way, a very clear image of a slice of sample is built up; if the sample is then successively moved short distances away from the microscope, with each time another slice being imaged, then a complete three-dimensional picture of the sample can be built up, which can be visualised and manipulated using a computer.

The major difficulty in fluorescence microscopy is finding the appropriate fluorescent dye and sticking it to the molecule one is interested in. The most common dyes are smallish organic molecules of the general type that we will discuss in more detail when we come to molecular electronics. At the moment there is some excitement about using tiny inorganic semiconductor particles a few nanometres in size—quantum dots. The advantage of these is that quantum effects mean that by controlling the size of the particle one controls the colour of the fluorescence. Both dyes and quantum dots have the disadvantage from the point of view of biological studies that, if you want to study the processes inside a living cell then you have to get them inside the cell without killing it. Another approach uses the fact that some organisms are naturally fluorescent. The favourite is a deep-sea jellyfish which has a creepy green glow. This comes from a fluorescent protein (imaginatively named green fluorescent protein, or GFP), and genetic engineering has been used to make all kinds of organisms produce variants of GFP attached to their own proteins, essentially making a living label.

Other types of waves

Powerful as light microscopes are, the diffraction limit means that they are fundamentally unable to look properly into the nanoworld. If you are going to use a microscope that depends on waves to form its image, then a proper nano-microscope is going to need waves whose wavelength is smaller than a nanometre. What kind of waves could we use?

Light, like radio waves, is an electromagnetic wave, which is based on oscillations in the electric and magnetic fields. These oscillations can take place at any wavelength, different wavelengths occupying different positions on the electromagnetic spectrum. Waves with wavelengths in the range of metres are the radio waves, which surround us all the time with a low-energy background of bad pop music. As we have seen, visible light has wavelengths in the range 400–700 nm, according to its colour. Is there a type of electromagnetic radiation, with a wavelength of a nanometre or less, which would allow us to make a microscope whose diffraction limit was low enough to have a clear look at the nanoworld?

There is such a radiation—these are X-rays. The wavelength of X-rays is exactly right, but, frustratingly, it is not easily possible to make a microscope that uses them for the very practical reason that it is very difficult to make a lens that will focus them. As we all know, X-rays easily penetrate most materials, and it is not possible to find a material that will bend X-rays without absorbing them. The X-ray pictures that we are familiar with, if we are unlucky enough to have broken a bone, are not really images in the true sense; they are simply unmagnified shadows. We will see below that X-rays have a vital role in allowing us to make sense of the nanoworld, but we cannot easily use them to make direct images.

So there is no type of electromagnetic radiation that we can use to make a nanoscope. But electromagnetic waves are not the only waves in the world;

quantum mechanics tells us that anything that behaves like a particle can also behave like a wave. The components of matter—protons, neutrons, and electrons, for example—can, in the right circumstances, behave exactly like a wave, and we can use them to make an image. Of these, by far the most useful is the electron.

The electron microscope

Some people say that electron microscopists are born and not made, and that is certainly a sentiment I can sympathise with. When I began my Ph.D. project in experimental polymer physics, as is usual, I was given a suggestion of an experiment to start things off. My supervisor, Jacob Klein, was interested in how polymers move around. A friend of his, a brilliant polymer chemist called Lew Fetters, had made some polymers that were not, as normal, linear, string-like molecules. Instead, they were shaped like stars, with as many as eighteen arms coming out from a central point. How could these move? They could not wriggle snake-like along their length, as the theory of reptation which I described earlier predicted for linear polymers. We would try and have a look, using what was then the state-of-the-art electron microscope at the Cavendish Laboratory, Cambridge. This microscope had the resolving power to see individual atoms, and it could distinguish between different chemical elements on the basis of the way the electrons lost energy. The way that Fetters had made the polymers meant that they each had a little knot of silicon atoms at the centre of the star, the rest of which was made of the carbon and hydrogen atoms that are typical in polymers. Could we see the knot of silicon at each molecule's heart, and thus have a way of tracking their motion?

With the blessing and advice of Mick Brown, the friendly Canadian professor whose machine it was, I sat down for the day in a darkened room that was completely filled by the giant microscope. With me was an experienced researcher who was going to begin to show me how to use it (of course, it would take a lot longer than a day for me to be able to use it by myself). In front of me was a bewildering array of knobs and dials, controls for the electron source, lens controls, beam steerers, apertures, and, at the centre of attention, a rather dim green television screen. I was completely left behind as the researcher tried all sorts of combinations of settings, but the result was always the same. The only thing we could see on the screen was the hole that the intense beam of electrons had physically punched through my delicate sample. Inexperienced though I was, I had resolved by the end of the day that I was not going to be an electron microscopist. I found something else to write a thesis about and the problem of how the stars get about was solved in another way.

Electrons are good for microscopes because they are very easy to handle. They can be made fairly straightforwardly; a filament, pretty much the same as that in a light bulb, will emit them readily when it is heated up. Because they are charged, they can be speeded up just by putting them through an electric field, and they can be steered in the same way. In fact, this is exactly what happens in a television set; a beam of electrons is produced, accelerated between two plates with a high voltage between them, and steered by variable applied electric fields. The beam hits the screen, which is coated with a material that glows when hit by electrons; in a television, the image is made by rapidly scanning the spot in lines across the screen. Electrons can be bent by magnetic fields, and this is how you can make a lens for them. A carefully-designed electromagnet can focus an electron beam, and it is these magnetic lenses that make electron microscopes possible. The precise control that electric and magnetic fields give you over the paths that electrons take is what makes electron microscopes harder to use than light microscopes; rather than a physical piece of glass, it is a pattern of magnetic fields that bends the electrons into focus, and adjusting these fields takes knowledge and experience.

The wavelength of electrons depends, according to the laws of quantum mechanics, on how fast they are going. The speed at which the electrons move depends, in turn, on how large an electric field you accelerate them through, which depends on the voltage you apply. A standard laboratory electron microscope will accelerate electrons through a few tens of thousands of volts, not that much more than the voltage applied to the electrons in your television set. Fast electrons have a shorter wavelength, and thus, in principle, would allow you to look at smaller objects; a high-resolution electron microscope, capable of resolving individual atoms, might use an accelerating voltage of a million volts or more. The apparatus needed to generate such a high voltage and keep it insulated is expensive and takes up a lot of space.

There are two ways of using electrons to form an image of a sample. The first is directly analogous to an ordinary light microscope; a sample, in the form of a very thin film, is illuminated by a broad beam of electrons, and then an image is formed from the electrons that have passed through the sample on a screen, using a pair of lenses. Such an instrument is called a transmission electron microscope. In the second method, lenses are used to focus the beam to a very fine point; this beam is scanned across the sample and the scattered electrons are detected. An image is built up by recording the intensity of the scattered electrons as the beam is scanned. In the most common type of scanning electron microscope, the beam is scanned across the surface of a sample, and electrons that have been scattered backwards off the sample are detected. But it is also possible to scan an electron beam across a very thin sample and detect electrons that have passed through it. This kind of instrument—a scanning transmission electron microscope—is capable of a very high resolution, and also, by looking at the energy that has been lost by the electrons as they pass through, can also do a chemical analysis of the tiny amount of material in the electron spot (it was this kind of microscope that I had such trouble with for my polymer sample).

If electron microscopes can potentially see down to the atomic level, what are the snags? The problems start from the fact that, because electrons are charged particles, if they collide with any kind of matter then they bounce off.

This means, in the first place, that the inside of the microscope needs to be pumped out to a high vacuum. This at once makes it impossible to look at any sample that contains water[2] because, of course, as soon as liquid water is put into a vacuum it rapidly boils off.

Also, electrons can only get through very thin films, so the sample has to be sliced very thinly—generally thinner than a micron, and much thinner than this if the best resolution is required. For soft materials like polymers, or for biological tissue, this is generally done by embedding it in epoxy resin and slicing it with an instrument called an ultramicrotome—a very high technology version of a kitchen mandolin, with a blade made of diamond rather than steel. Biological tissue first needs to be 'fixed' and 'stained'—chemicals are added which harden up the soft macromolecules and attach heavy metals to them to make them more visible to the electrons. Hard materials like semiconductors or metals need to be thinned by being bombarded by ion beams. After all this rigmarole, it is not always easy to be certain that what you see in the microscope actually has much relationship to what you started with.

Another potential difficulty is that electrons can physically and chemically damage your sample. After all, what you are talking about is a form of ionising radiation (the beta particles emitted in radioactive decay are nothing other than energetic electrons), and the levels of radiation that the sample is seeing would be enough to kill any organism. The problem is worse in that, to achieve the highest resolution, one needs to use very bright sources of electrons and focus them onto a very fine point, resulting in a beam so energetic that if you are not careful you can physically punch a hole through your sample (this is the problem I was having with my polymer).

Despite these difficulties, electron microscopy was the technique that first allowed us to see the nanoworld. In materials like metals, it made it possible to see directly how the atoms were arranged in their crystalline lattices, and, perhaps even more importantly, to see the imperfections in the structure that actually determine the properties of real materials, rather than the theoretical idealisations. In biology, the existence of nanoscale structures within cells—organelles—was revealed, as well as the structures of the smallest of life-forms, viruses.

Electron microscopy continues to develop, but perhaps the most exciting recent advances have come, not so much from improving the hardware, but from devising ways of preparing samples that minimise the disturbance they are put through. Ways have been found of very rapidly freezing biological material, with rates of cooling being achieved that are so fast that the water does not have a chance to form normal, crystalline ice. The growth of ice crystals, even on the tiniest scales, destroys the fine structure of biological material, as anyone who

---

[2] Recently, a scanning electron microscope has been developed that does allow you to do this— the environmental scanning electron microscope. This works by separating a sample chamber at a relatively high pressure (though still far from atmospheric pressure) from a column pumped out to a very good vacuum with a very small aperture, through which gas leaks more slowly than it is pumped out.

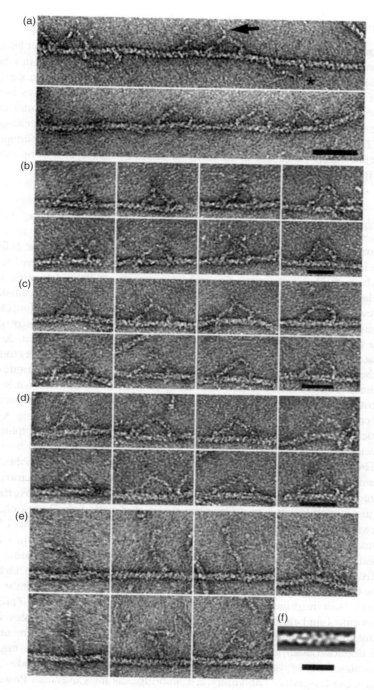

Fig. 2.2 Cryo-transmission electron microscope images of a motor protein walking along a track. From M. L. Walker, S. A. Burgess, J. R. Sellers, F. Wang, J. A. Hammer III, J. Trinick, and P. J. Knight. *Nature* **405** (2000) 804–7 with permission from the copyright holder, Nature Publishing Group.

has compared the texture of frozen strawberries to that of the fresh fruit knows. But on very rapid cooling, the water forms, not a crystal, but a glass, in which the molecules are frozen in the same disordered positions they had in the liquid. If samples frozen like this are kept at these ultra-low temperatures in the electron microscope as they are being imaged, then one has the hope of seeing the nanostructures just as they are in life, frozen in a moment. This technique, cryo-TEM, has recently produced some remarkable pictures of biological molecular machines caught in the different stages of their cycles of operation (see figure 2.2).

Scattering and diffraction

The structure of DNA, the molecule that carries the information that defines all living organisms, was discovered almost exactly fifty years ago as I write this. It is an iconic structure, two polymer strands circling each other in a double helix, linked by pairs of hydrogen-bonding bases, like a twisted ladder. However, the structure was not discovered by the direct visualisation of an image from a microscope; it was inferred by studying the patterns of dots made by a beam of X-rays scattered from a crystal of the material. X-ray diffraction is the technique which, more than any other, has allowed us to find out the structure of the nanoworld. The structure of DNA, the structures of proteins, and increasingly now the complex structures of the assemblies of proteins that make up biological nano-machines—all of these discoveries, now codified in huge, publicly accessible databases, were made using X-ray diffraction, the technique that almost by itself produced the new discipline of molecular biology.

The phenomenon of X-ray diffraction was discovered in 1912 by Max von Laue at Munich University. But it was a young graduate student, Lawrence Bragg, who discovered a simple way to interpret Laue's photographs. Bragg had just started his studies at Cambridge, as a student of J. J. Thompson, the discoverer of the electron, and was almost embarrassed at the simplicity of the method he had discovered. But it laid the foundation for a career in which he and his protégés transformed both chemistry and biology. It turned into a family business; Lawrence's father, William Bragg, was Professor of Physics at the University of Leeds, and within a year William had constructed the first purpose-built instrument for making X-ray diffraction measurements. Through the twenties and thirties, the structures of more and more molecules were discovered; first, the relatively simple crystal structures of inorganic materials like diamond and simple salts, then increasingly complicated organic molecules. In 1938, Lawrence Bragg returned to the scene of his student discovery, the Cavendish Laboratory at Cambridge, as the Cavendish Professor and Head of the Laboratory. There, his student, Max Perutz, succeeded in getting a diffraction pattern from haemoglobin. Meanwhile, in London, Desmond Bernal (who was almost as famous for his ardent communist views as for his science) had obtained a diffraction pattern for the protein lysozyme. But the diffraction patterns—the regular-looking arrays of spots on the photographic films—for these big molecules were very complex. The ability of the experimentalists to obtain good diffraction patterns had temporarily outrun the ability of the theorists to devise tractable ways of interpreting them, and it was not yet possible to solve the problem of the structure of these proteins.

The breakthrough decade, in which X-ray crystallography turned into molecular biology, was the 1950s. The most famous breakthrough was in 1953; Rosalind Franklin, working in loose association with Maurice Wilkins at King's College, London, had obtained a particularly clear diffraction pattern from DNA. Francis Crick and James Watson, a pair of brash young researchers in the Cavendish Laboratory, solved the structure. The question of the proper division of credit between these four (and indeed with Ray Gosling, the graduate student at King's who actually prepared the crystal) has been controversial ever since, not least because Rosalind Franklin had tragically died before Wilkins, Crick, and Watson were awarded a Nobel prize. Much less famous, but possibly no less important, was the first solution of the structure of a protein. In 1957, John Kendrew, also working at Cambridge, solved the structure of myoglobin. Proteins have a much less regular structure than DNA, and their diffraction patterns are correspondingly more complicated; this advance opened the way to the discovery of the structures of the 10 000 or so proteins whose atomic coordinates are now known and stored in protein databases.

How does X-ray diffraction work? Again, it depends on the fact that X-rays are a wave. If an X-ray encounters a single atom, then it will be scattered. Absorbing some of the energy of the wave, the atom will re-radiate it in all directions. If two atoms are sitting side by side, then they will each scatter the X-ray, but the waves scattered by each atom will interfere with each other. In some directions, where the wave scattered by one atom has a peak, the wave scattered by the other atom has a trough. In this direction, the peaks and troughs of the waves scattered by the two atoms will cancel each other out, and a detector mounted there would see a blank spot—no X-rays would be detected. In other directions, the peaks and troughs will reinforce each other, and we would see a particularly strong X-ray signal. If we measure the X-ray intensity (or if we look at an exposed X-ray film) then we will see a pattern of alternately bright and dim stripes. If we know the wavelength of the X-rays, then we could deduce from this pattern how far apart the atoms were.

A protein crystal has, not just a pair of atoms, but a three-dimensional array of molecules, each of which has many hundreds of atoms. X-rays are scattered off each of the atoms in every molecule, each different kind of atom scattering by a different amount. All of these scattered X-rays interfere, peaks and troughs sometimes adding, sometimes partially or completely cancelling each other out. The result is a complicated pattern of spots, each with a different intensity. It is this pattern of spots, some bright, some less bright, that needs to be interpreted to determine the structure.

The mathematical relationship between the diffraction pattern and the structure is a complicated one, and there is a fundamental problem with it. If you know a structure, then you can with complete confidence calculate the diffraction pattern, but the same is not true in reverse—in principle, a number of different structures could produce the same diffraction pattern. To overcome this problem requires a combination of intuition and a number of subtle mathematical and experimental tricks.

In the days before computers, the calculations were long and difficult, but now life is much easier. Very bright sources of X-rays are available; particularly from machines that accelerate electrons to great speeds, spinning them round in rings, tens of metres in diameter, at velocities close to the speed of light. These electron synchrotrons were originally built for particle physics experiments, but their importance for structural studies, particularly of proteins, is so great that accelerators are now built expressly for this purpose. Fast computers have almost completely automated the process of solving the diffraction pattern to find the structure, and easy-to-use computer graphics programs make it easy to visualise the three-dimensional arrangement of atoms once it has been discovered.

The only thing that remains difficult, and the province of old-fashioned craft skills, is the growing of the crystals in the first place. Proteins are big, soft molecules which do not have a particularly strong driving force to pack in a large, well-ordered crystalline array, in the way that hard spherical objects might. One much talked about way of improving the situation is to grow the crystals in space, where the delicate ordering process would not be perturbed by gravity. Unfortunately, there is no real evidence that protein crystals grown in space are actually significantly better than those grown on Earth, despite their huge expense.

Imaging versus scattering

It is my belief that scientists who use microscopy and those who use diffraction and scattering belong to two quite different tribes who do not quite speak the same language, and who certainly do not entirely trust each other. The complicated mathematical transforms that relate a structure to a scattering pattern eventually become intuitive to a diffraction scientist, but to outsiders they remain more than slightly mysterious. I once heard a distinguished scientist say that he could never entirely believe the results of a scattering experiment, because it was like trying to guess what was in a darkened room by standing outside and throwing balls into it. For such people, seeing the image from the microscope is believing. But the apparent immediacy of a microscope image can be misleading, too. Aside from all the potential difficulties that arise from the way samples are prepared, there is the fundamental point that in an image you are seeing only one example of what you are looking at, whereas a diffraction pattern contains information averaged over all the molecules your beam is illuminating. It is a well-known joke amongst the more cynically inclined scientists that, when you see a picture reproduced in a scientific paper showing a beautifully clear image of some complicated nanoscale structure, and the picture is referred to in words rather like this: 'Figure 3 shows a typical image . . . ', then you know that the image is anything but typical—it has been carefully selected from many much less clear pictures. This does underline the fact that statistical significance is always a potential problem when you are looking at only one example of a structure at a time. Both scatterers and microscopists, and all who use their results, should remember that there are many working assumptions and complex manipulations between the reality and what they see in their pictures or deduce from their diffraction patterns.

Scanning probe microscopy

Often, in the history of science, the introduction of new techniques can be as important as the development of new theories or concepts in crystallising a new field. For example, the new field of molecular biology was created by the technical development of X-ray diffraction to the point at which it could be usefully applied to biological molecules, rather than by any theoretical advance. What has catalysed the emergence of nanoscale science and technology as a discrete discipline was the invention of a group of related techniques known as scanning probe microscopies. First to be invented was scanning tunnelling microscopy, created by Binnig and Rohrer in 1981; this was shortly followed by the invention of the atomic force microscope by Binnig, Quate, and Gerber in 1986.

These instruments produce images of surfaces at a resolution that can discern individual atoms and molecules. But they way they work is very different from the mode of operation of either light microscopes or electron microscopes. Rather than visualising the sample using a wave, what you are doing with a scanning probe microscope is much more like feeling your way across the surface. In a scanning tunnelling microscope, a very fine metal needle is scanned across the surface. By measuring the electrical current that flows between the surface and the needle, and continually adjusting the height of the needle above the surface so that the current is constant, one can build up a picture of the surface in just the same way that one would discover the contours of a rough surface in a dark room by running ones fingers across it. In fact, this analogy applies even more closely to the atomic force microscope. Here the interaction between the surface and the probe is not mediated by an electrical current, but by the forces that occur between objects on very small scales. The probe of an atomic force microscope consists of a tiny pyramid-shaped tip that is mounted on the end of a thin, silicon cantilever. This acts as a very weak spring; as the tip is brought close to the surface, if the surface attracts the tip then the cantilever will bend. This bending is detected simply by reflecting a laser beam from the end of the lever and measuring the deflection of the reflected spot.

What is needed in order to make scanning probe microscopes work? Firstly, one needs to be able to move the tip around with great precision and accuracy. This is achieved by using piezo-electric materials. A piezo-electric crystal is a material that, when one deforms it slightly, develops an electric field across its faces. The reverse of this property also holds; if you apply a small voltage across the crystal then it will slightly contract or expand. The most well-known piezo-electric material is quartz, and the use of quartz crystals in watches depends on this property. In scanning probe microscopes, a material called lead zirconium titanate (known in the trade as PZT) is used to make scanners, which respond with motions on the nanometre scale when small voltages are applied. This allows us to move the tip around with great precision, allowing us both to scan across the sample and to move toward and away from the sample surface.

Very sharp tips are also essential for resolving very small details. The kinds of sharp objects that we are familiar with in the house, like pins and needles, are, if you magnify them enough, not that sharp at all, even on the micron scale, let alone on the nanometre scale that we are going to need if we are to be able to use such a tip to probe individual atoms. So the tips that are used for scanning tunnelling microscopes and for atomic force microscopes need to be made using special techniques such as microlithography, as discussed in the next chapter. Because they are so sharp, they are also rather delicate, and they do not last very long.

In the scanning tunnelling microscope, a handy piece of physics makes the job of making a sharp enough tip less difficult than one might at first imagine. If the tip is held at some very small distance away from the surface—a distance measured in nanometres or less—then the way in which electrons flow across this empty gap is by the quantum mechanical process known as tunnelling. On the macroscopic scale, electrons do not spontaneously flow out of a metallic conductor and jump across a gap to another conductor. This is just as well; otherwise you would get an electric shock just by walking underneath an overhead power line. But things are different at the nanoscale. According to quantum mechanics, an electron can briefly enter a region that classical physics would forbid it from going into. But the probability that it does this—that it can leave the conducting metal of the tip, tunnel through empty space, and arrive at the surface—is an exponentially decreasing function of the distance across the gap. This means that, even if the tip is relatively blunt, it is the part of the tip that is closest to the surface that contributes by far the most to the current you measure.

Even the best-prepared tips sometimes do not turn out as planned, and one has to be on one's guard in scanning probe microscopy for images that are not what they seem, because the tip is not the ideal, single sharp point. For example, if, instead of having one sharp point, there are two closely-spaced points, then a single feature would appear in the image doubled. Being able to recognise this kind of tip artefact is just one of the skills a scanning probe microscope operator needs to have, and it is this sort of detail that makes the use of scanning probe microscopes not entirely routine.

Another feature, that while not absolutely essential, has made scanning probe microscopes very much easier to use, has been the availability of enough cheap computing power to be able to generate and process attractive images from the raw data that the instrument supplies. What the instrument actually produces is a trace of the voltage as a function of time as the tip is moved across the surface. While it would be possible to process this to produce a picture using analogue electronics, in much the same way as a television produces a picture by flicking a spot in lines across a screen, it is much easier and more flexible to do the whole job digitally. This has the substantial technical advantage that one can digitally treat the image to remove experimental artefacts (and the aesthetic advantage that the resulting image can be presented in the most attractive set of colours).

In favourable circumstances, molecules and even atoms can be imaged with scanning probe microscopes. These instruments were not the first to permit one to look at the molecular world; this distinction belongs to the electron microscope. Nor have they come close to matching the precision of X-ray diffraction in determining the structures of biological molecular machines. So why have scanning probe microscopes been so important in catalysing the emergence of nanoscale science and technology as a separate field?

What really sets the scanning probe microscopies (and particularly atomic force microscopy, see figure 2.3) ahead of competitor techniques is their low cost, their ease of use, and the fact that one can look at samples without much in the way of sample preparation. A research-grade atomic force microscope (AFM) might cost in the region of £100 000—a lot of money, certainly, but it is substantially less than the cost of the high-resolution transmission electron microscope that would be needed to achieve a similar resolution. I would not

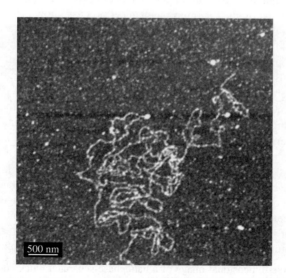

Fig. 2.3 Atomic force microscope image of a single DNA molecule. Reprinted from *Surface Science* **491**, A. D. L. Humphris, A. N. Round, and M. J. Miles, Enhanced imaging of DNA via active quality factor control, p 468–472 Copyright (2001), with permission from Elsevier.

try to argue that an AFM is completely straightforward to use, but a graduate student might be able to start getting useful results after a week or so of training. Entry into the freemasonry of electron microscopists takes a much more painful period of initiation. But, most important is the fact that it takes much less fuss to prepare the samples to look at in an AFM.

There is as much skill—perhaps more—in knowing how to prepare and present a sample for a transmission electron microscope as there is in knowing how to get the best results from the instrument once the sample is inside. As we have seen, there is the fact that it is only possible to look at very thin slices and there is the difficulty of having to put the sample in a vacuum. But an AFM can operate quite happily underwater; as long as it has a surface to look at there is no need to make a thin section. Using an electron microscope is a big undertaking, but you can usually simply slip your sample from the laboratory into an atomic force microscope with very little ceremony.

One shortcoming of the scanning probe microscope is that it is restricted to studying the surface of things. The sample preparation for an electron microscope is difficult, but it is at least possible to make many thin sections of one's material using an ultramicrotome. Having prepared a stack of sections, rather like a set of slices of cucumber, one can image each section sequentially and then use the set of images to reconstruct the original three-dimensional structure of the sample. In fact, it is possible to do something like this with an atomic force microscope; rather than using a knife to take sections off the sample, one can instead use a plasma to etch away a few nanometres from the surface. Each new surface thus freshly revealed can be imaged, and the resulting stack of images put together to give the three-dimensional structure (see figure 2.4). At

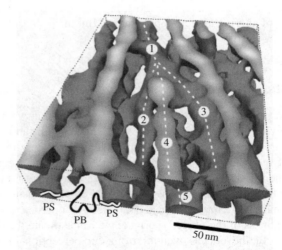

Fig. 2.4 A three-dimensional image of a nanophase-separated block copolymer morphology imaged by a combination of AFM and plasma etching. Reprinted with permission from R. Magerle. *Physical Review Letters* **85** (2000) 2749, Copyright 2000 by the American Physical Society.

the moment, this method is too labour intensive to be widely used, but it may be possible to automate it.

One final advantage of scanning probe microscopes is that, just as we can use our fingers to pick objects up and move them around as well as to touch them and feel their shape and outline, so can we use a scanning probe microscope to manipulate the nanoworld. We can scratch lines and draw more complicated shapes on surfaces, and we can pick up individual molecules from one place and move them elsewhere. We can pull molecules and stretch them, measuring their elastic properties one by one. We can inject an electron into them, measuring the electrical properties of a single molecule. We can even carry out a chemical reaction at the point of the tip, literally building up a chemical structure atom by atom.

Actually, all microscopes can in principle modify what they look at. Even a light microscope, if the illumination is strong enough (and this is particularly the case in confocal microscopy), can produce enough light to bleach what you are looking at, just as dark hair can be bleached by bright tropical sunlight. Electron microscopes are notorious for producing damage in delicate samples— after all, what we are subjecting our sample to here is a massive dose of radiation, sufficient always to cause massive chemical damage, or even the complete vaporisation of the material. This principle is used for nanofabrication techniques like electron beam lithography. In this technique, a sharply-focused electron beam is used in what is, in effect, a modified scanning electron microscope, to write patterns directly into a material that is specially designed to be easily damaged by the radiation of an electron beam. But it is the use of scanning probe microscopes to modify what you are looking at that has caught people's imagination the most, because they do offer the tantalising possibility of altering matter atom by atom. But this subject belongs to the next chapter.

Living in the nanoworld

Our acute and highly-developed sense of vision means that we are very well adapted for living in the macroworld. We can almost instantly take in the details of a three-dimensional scene, appreciating intuitively the spatial relationships between objects. We can deduce quite a lot about the properties of things from their surface appearance (shiny, dull, colour, etc.). For things we can reach out to or pick up, we can supplement our vision using our sense of touch, to get much more information about material properties. We can feel the degree of roughness of the object's surface; we can estimate the density from our perception of the objects weight; from our perception of whether the object feels hot or cold we make deductions about its thermal properties, which we can interpret to distinguish whether something is made from metal or plastic, for example. If we can pick up the object, we can rotate it to look at it from various angles, to deduce something about its internal structure. But, apart from those few objects that are made of transparent materials, we cannot yet see inside things. Doctors need to use sophisticated tools—in medicine we need X-rays and magnetic resonance imaging to look inside someone without cutting them open.

The way our brains process the outputs of our senses of vision and touch provides the rich and intuitive sensory environment that gives us our sense that we fully inhabit the macroworld, and allows us to interact with and manipulate objects in that macroworld. Is it possible to extend our senses in such a way that we can construct a similarly rich sensory environment for interacting with the nanoworld?

To construct such an environment, we would need two things: an interface to our human senses that is as unobtrusive as possible, and input to that interface that provides information about the nanoworld that is as rich as the information our senses provide about the macroworld. Our interface with the nanoworld at the moment is simply the computer screen on which the image from an atomic force microscope is displayed. This gets the job done, but does not give us much sense of being immersed in the nanoworld. It is the same kind of interface that the most basic sort of computer game has, like the original flight simulation games in which you saw a crudely rendered view out of the cockpit on the screen, and you attempted to steer the aircraft while trying to remember whether it was the letter 'g' or 'h' that turned the rudder left. The sort of flight simulator that real pilots train on fully immerses you in the simulated world. A chair in front of a single computer screen is replaced by a realistic replica of a cockpit, images are displayed on every window, the controls have the same mechanical response as the real thing and even the motion of the cockpit is reproduced in the simulator. This degree of sensory immersion gives the trainee pilot the strong illusion of flying a real aircraft. We can anticipate a similar virtual reality interface being constructed for a journey into the nanoworld.

Where would the input for this recreation of the nanoworld come from? It would be quite possible to use a computer simulation, just as a flight simulator does. A library of micrographs, either from an electron microscope or an atomic force microscope, would be used to construct the scene to be explored. A computer simulation, written to accurately capture the physics of the nanoworld, would generate the rules expressing how objects moved around, and interacted with each other and with the human participant. Such a simulation would be entirely technically feasible now, though it would not do much more than provide an educational and entertaining way of learning about the nanoworld.

What would be more impressive would be if the virtual world generated inside the computer and conveyed to the participant through the immersive interface actually reflected reality. A ground-based pilot can remotely fly a pilotless aircraft; images from the aircraft, as well as readings from the instruments, are received by the ground station and used to recreate the environment of a virtual cockpit. Similarly, we could imagine images from microscopes being used to construct a virtual rendition of a real nanoscale scene, with which a human operator could interact remotely.

In fact, the first steps to this vision have already been taken; some kinds of virtual reality interfaces to atomic force microscopes have already been built. What is needed to create a fully-immersive environment with which to explore and interact with a real nanoscale environment? Atomic force microscopy is clearly the first choice as the basic visualisation technique; this can make observations in the nanoworld without the need for sample preparation, and thus offers the possibility of making dynamic observations, showing how things change in real time. Some development of the technique is needed to increase the rapidity with which a picture can be taken; current-generation AFMs have scan speeds that are too slow to record videos in real time, but work is going on to improve this. One limitation of AFM is that it can only satisfactorily image things in close proximity to a solid surface. Ideally, one would wish to image matter that floated freely in solution with a resolution of nanometres or tens of nanometres. A microscope that could achieve this would be a boon for disentangling the complexity of cell biology. Alas, at the moment it is not at all clear how such an instrument could be built.

3
Nanofabrication

Introduction

I do not like to think of myself as old, but, in the time it has taken me to grow up, an industry which dominates our world has itself grown to maturity from virtually nothing. That industry is of course the personal computer industry, and the technology on which it is based, in its modern form, is the integrated circuit. When I was fourteen or so, I was being taught to do calculations with a strange analogue device called a slide rule, while one of the richest and most cosmopolitan of my fellow pupils amazed the rest of the class with his new, and very expensive, pocket electronic calculator. By the time that I had left school, I had seen my first personal computer, though of course such devices were far too new-fangled to play any part in my undergraduate education (I went to a very ancient and conservative university). During the time I did my Ph.D., IBM launched their Personal Computer, bringing corporate respectability to what had previously been the domain of hobbyists, and pretty much every aspect of the modern world changed for ever.

What made all this possible was the relentless shrinking of the components on the integrated circuits that form the heart of a modern computer. An integrated circuit is a circuit made up from a number of electronic components—most importantly, transistors—made on the surface of a single monolithic piece of the semiconductor silicon. What has made possible the phenomenal growth in computing power has been the rapid progress in the size of the features that it is possible to define on the surface of the chip. Smaller feature sizes have a dual benefit; the smaller the feature size the more components can be fitted onto a single circuit, but, in addition, the smaller the component the faster it can be made to run. The first microprocessor—the Intel 4004—contained 2300 transistors on a 3 mm by 4 mm chip, with 8 μm features, and was made in 1971. Thirty years later, Intel's flagship microprocessor was the Pentium 4. This was somewhat bigger physically, with an area of about 200 mm^2, but it contained an astonishing forty-two million components, with line-widths of 0.18 μm. In electronics, we are well along the transition from microtechnology to nanotechnology.

How small does a man-made structure need to be before we can count it as nanotechnology? A (successful) candidate for a job at my university was asked this question; he replied with firm authority, '100 nanometres'. 'Why 100?' the panel asked, to which he replied without a moment's hesitation, 'Because President Clinton says so.' This seemed a very fair answer—100 nm was indeed the figure associated with the USA's National Nanotechnology Initiative, and the invocation of the authority of the US president pretty much sums up the arbitrary nature of definitions like this.

But 100 nm is an interesting threshold for another reason besides its association with one of the western world's more colourful recent political figures. The globally massive semiconductor industry has made the transition to volume production of integrated circuits with a minimum feature size of less than 100 nm. In this sense, nanotechnology has already become a mainstream, hugely valuable global industry.

These extraordinary advances have arisen from the rapid perfection of a single manufacturing technique, which now is able to make devices approaching the nanometre scale. The essence of this 'top-down' approach to making tiny devices is the combination of an ability to deposit very thin layers on a flat surface, with methods to etch away selective regions that are defined by masks. Conceptually, the method is not that different to the technique used by artists to make the copper engraving plates with which old books are illustrated. These engravers would take a flat copper plate, coat it with a varnish, and then make their drawing by scratching through the varnish to the copper. The plate would then be exposed to an acidic etching solution, which would dissolve away the copper where the varnish had been scraped away. In another variant of the technique—relief etching—varnish is directly painted onto the copper; when the copper is etched the material around the line is dissolved away, leaving the line in relief, standing about one-tenth of a millimetre above the rest of the copper. This is the technique used to stunning effect by William Blake to create his illuminated books, such as *Jerusalem*.

Artists can create lines which are fractions of a millimetre wide, and etch the copper away to similar depths. Today's engineers can make a line much less than 100 nm wide. Combine this process of coating a surface with a protective layer—called a resist in the micro-electronics business, patterning the resist to allow a selective etching step—with other techniques for depositing very thin layers of material, and one has a recipe for creating complex structures at very small sizes.

The transistor

The electronics industry is the largest industry in the world, but to a large extent it depends on just one invention—the transistor—and one way to make and combine transistors—what is called 'complementary metal oxide semiconductor' technology, or CMOS.

The predecessor of the transistor was the thermionic valve, and the word valve conveys quite well the essence of what a transistor does. The transistor has two terminals connected by a conducting channel; if you connected the terminals up to a battery, then a current would flow. A third terminal, called the gate, controls the amount of current the transistor can carry. In the type of transistor called a field-effect transistor, a voltage applied to the gate has the effect of changing the number of charge carriers in the channel; the more charge carriers there are the more current the channel will carry for a given voltage across it. It is like a tap; the mains pressure forces water through the tap, but we can control how much water flows for a given pressure by turning the tap on and off.

Semiconductors have the appealing characteristic that we can controllably introduce charge carriers into them by adding small numbers of impurity atoms, a process called doping. One type of impurity has the effect of introducing what are effectively free electrons—particles of negative charge. A semiconductor doped in this way is called n-type. Another type of impurity produces the effect of positive charges; this produces a p-type semiconductor. Actually, a more accurate way of thinking about this is that a p-type semiconductor has a deficiency of electrons. Think of one of those puzzles with sliding tiles, in which one successively moves the tiles into the one empty space; the effect of moving tiles into the space is that space itself effectively moves around. In the same way, in a p-type semiconductor, it is the absence of an electron—an electron vacancy or hole—that effectively carries the charge, rather than the presence of a true, physically-existing positively-charged particle.

We can have two types of field-effect transistors, then. In an n-type transistor, the conducting channel is made up of an n-type semiconductor, in which the charge carriers are electrons. If we put a positive voltage on the gate, then this has the effect of removing charge carriers, effectively shutting the valve and preventing any current flowing. Thinking of the transistor as a switch rather than a valve, a positive signal switches it off. The reverse is true in a p-type transistor, where it is not electrons, but holes, that carry the current. If we apply a positive voltage to the gate of the p-type resistor then we withdraw even more electrons, creating even more holes, and the channel conducts. We would need to apply a negative voltage to the gate to turn off the p-type transistor.

Digital logic is carried out by assigning logical values to voltages. A voltage of zero corresponds to a logical value of zero (i.e. false), while a fixed positive voltage corresponds to logical value of one (i.e. true). Digital circuits are designed so that each transistor operates as a switch. Imagine the two types of transistors as being two types of switches, one of which is switched on when the button is pressed, one of which is switched off. The basic idea of an electronic computer is that a combination of these two kinds of switch can carry out any logical computation.

The simplest logic circuit is one that inverts its input; if the input is 'true', then the output is 'false', and vice versa. This can be made with one transistor of each type. A circuit is made between a conductor at a positive voltage and a conductor at zero voltage through one transistor of each type, linked in series. Between the two transistors is the output. The same input is applied to the gates of both transistors. If the input voltage is high, representing logical one, then the n-type transistor is turned off and the p-type transistor is turned on. The output is directly connected through the p-type transistor, which is switched on, to the conductor at zero voltage—it reads logical zero. If the input voltage is zero, representing logical zero, then the reverse happens; the output is connected through the switched-on n-type transistor to the high-voltage conductor, and it reads logical one.

Inversion is the simplest logical operation, and it can be carried out with just two transistors, one of each type. More complex logic takes two input values to compute a single output. For example, the Not-OR function (NOR) returns one if both inputs are zero, and zero for all other combinations of input. This function can be implemented with four transistors, two of each type. Remarkably, it can be shown that any logical computation can be created by linking together these two functions, the inverter and the NOR. We have one simple inverter circuit, and one simple NOR circuit, and combinations of many of these two circuits can execute any logical or numerical calculation.

The type of field-effect transistors that are used in today's computers are called metal oxide semiconductor field-effect transistors—MOSFETS, for short—because that succinctly describes their structure. The basic design principle which I have just described is known as CMOS—standing for complementary metal oxide semiconductor, because two complementary types of transistor are used. CMOS technology dominates in the design and manufacture of integrated circuits for computer memories and central processors.

It is remarkable that all of the powerful capabilities of modern computers—the graphics, the games, making calculations, editing videos, and writing words—all derive from the simple manipulation of ones and zeros. It is even more remarkable that these symbolic manipulations are carried out by only two basic components in a couple of simply-connected combinations. We will see that this is not the only way to design a calculating machine—our brains work in quite a different way. But it is the way that is most consistent with the whole tenor of modern manufacturing methods, the massive, relentless multiplication and connection of a few identical and interchangeable components.

Making integrated circuits

Integrated circuits are sometimes called silicon chips, and, not surprisingly, the starting-point for making one is a block of the element silicon. Silicon is not a rare material—its oxide, silica, is found in nature in its pure form as the mineral quartz, and it is the major component of most sands and of the rocks sandstone and quartzite. The element itself is a hard, shiny, almost black material, with a crystal structure identical to that of carbon in diamond. (The element silicon is not the same as the polymeric material referred to as silicone. This is a long-chain polymer, often cross-linked to form a rubber, whose backbone is made up of alternate silicon and oxygen atoms. Silicon is used for making integrated circuits, silicone for filling the gap round your bath and for breast implants. It is best not to confuse the two.)

The process of making an integrated circuit starts with making a large, single crystal of extremely pure silicon. This is cut up into slices a fraction of a millimetre thick, which are polished to a mirror finish. These are called silicon wafers; they are generally in the form of a circle up to 30 cm in diameter. On each of these featureless, smooth disks, billions of individual, tiny components will be created to make the integrated circuits that form the microprocessors and memory modules of a modern computer.

Only a few basic processes are needed to make the transformation from a mirror-like surface to these intricate circuits. These processes are the growth of thin films, the creation of patterns on the surface of these films, the selective removal of parts of the patterns by etching, and the introduction of impurity atoms near the surface. By combining and repeating these simple unit processes many times, structures of great intricacy, and tiny size, can be made in huge numbers.

What kind of thin layers do we need? Hard, insulating layers are needed to separate the conducting wires and plates that make up the transistors and other components. Semiconducting layers are needed to make up the individual components of the transistors. Metal wires connect up the different parts of the circuit. And finally, we will see that in order to pattern the layers horizontally you need to apply thin layers of a material that can be patterned horizontally and then selectively removed—these resist layers are made from polymers.

The most useful insulating material in silicon semiconductor technology is silicon dioxide—quartz. This has the great advantage that it can be grown directly on the surface of a silicon wafer, by oxidation of the underlying silicon in an oven at around 100 °C. Layer by layer, the silicon is transformed into its dense, hard oxide, in a reaction which can be controlled with great precision to yield a film with a thickness anywhere between a few nanometres up to a micron. Of course, this kind of thermal oxide can only be grown when it is needed directly adjacent to silicon; if we need to include an insulating layer in some other position in the complex multilayers that comprise an integrated circuit, then we have to use some other method.

One flexible method for making thin films of a variety of materials, including metals, semiconductors, and insulators, is known as chemical vapour deposition. The idea behind this is that one introduces a gaseous precursor compound to the material one wants to deposit, which decomposes on the surface of the wafer to leave a thin, uniform film of that material. For example, silicon forms a highly-reactive, gaseous compound with hydrogen called silane (SiH_4), which will react on a surface held around 650 °C, depositing a highly-uniform silicon film at rates of tens of nanometres a minute. Other precursor gases can be used to grow other useful materials, including insulators such as silicon dioxide and silicon nitride, and metals like tungsten and aluminium.

Metal layers can also be applied by simple evaporation. This process sounds unlikely—one simply heats up the metal to such a high temperature that it first melts, and then evaporates, to be condensed on the surface of the wafer. But it is simple to do for the very small quantities of metal that are required. Applying the thin polymer layers that are needed to act as resists is even simpler—the polymer is dissolved in an organic solvent to make a rather thin, glue-like dope. The wafer is mounted on a turntable, the dope is poured over it, and the turntable is made to spin at a few thousand revolutions per minute. Most of the polymer solution is flung off, leaving only a thin coating, which quickly dries to form a very uniform and smooth layer, whose thickness can be accurately controlled between a few nanometres and a micron or so.

Making layers of material whose thickness is measured in nanometres is actually relatively easy. Just from ingredients and equipment in your local supermarket you could probably make a uniform 20 nm layer of polymer on a flat substrate like a piece of glass. But how can we make something like a wire from a thin layer like this? We would need to pattern the layer horizontally, to define the edges of the wire, and then remove the excess material outside these limits. The dominant technique for horizontal patterning in today's semiconductor is photolithography. The progress in shrinking the size of components down into the realm of nanotechnology is directly related to a series of improvements in this technique. It is not at all clear how much further this process can go; it is possible that within the next ten years some fundamentally new technique will be needed if we are going to continue to make components ever smaller and computers ever more powerful.

The origins of photolithography are to be found in the early history of photography. As we have seen, the idea of a resist layer that would prevent the etching of a plate underneath it, transferring a pattern from the soft resist layer to a harder substrate, like a copper plate, was known to eighteenth-century engravers like William Blake. The next step was to devise a way of transferring a pattern to the resist by selectively illuminating it, rather than by directly scraping it away mechanically. The first photograph in the world was made using just such a direct process (see figure 3.1). Nicéphore Niépce had been experimenting for some time with methods to fix the transient image made by

Fig. 3.1 View from the window at Le Gras, Nicéphore Niépce. The world's first photograph, created in 1827 using a light-sensitive asphalt as a negative resist. From the Gernsheim Collection, reproduced by permission of the Harry Ransom Humanities Research Center, The University of Texas at Austin.

a camera obscura. Success came when he used Bitumen of Judea, a type of natural asphalt which hardens when it is exposed to light. He spread a thin layer of this material, dissolved in oil, onto a metal plate. The image of the camera obscura was allowed to fall onto the plate, and after being exposed all day in the sun the plate was removed. The unhardened parts of the bitumen were washed away with oil, leaving an image in which the highlights were defined by bitumen and the dark areas by bare metal. Niépce's first photograph, of the view out of the window of his mansion in northern France, was taken in 1826 and survives to this day.

Niépce went further, taking the obvious step of transferring the pattern of hardened bitumen to the metal plate beneath by etching it with acid. The resulting relief plate could be inked and prints taken from it. This technique became known as photoetching, but, in essence, it is identical to the technique of photolithography as it is used today.

In the modern process, it is more common to use a polymer that becomes easier, rather than harder, to dissolve when it is exposed to light. This is called a positive photoresist (Niépce's asphalt behaves like a negative photoresist). The first step is to create a mask—an image of the pattern that defines the structure of the circuits at that level. The mask is based on a fairly substantial chunk of silica glass, on which the pattern is drawn in a thin film of metal. The wafer is coated with a thin layer of a light-sensitive polymer—the resist—and an image of the mask is made on the wafer's surface. Where the resist is exposed to light, it becomes more soluble in solvents. So, by exposing the resist to such 'developers', the pattern on the mask is transferred to a pattern on the wafers of remaining areas of resist.

What is the lower limit in size of features that can be defined photolithographically? The issue is exactly the same one that limits the smallest size of feature that can be seen in an optical microscope—diffraction. This is the blurring that arises as a result of the wave nature of light—it means that even if the line that is defined on the mask is very thin indeed, the image of that line that is projected onto the wafer will have a certain width which is related to the wavelength of the light that is used to make the image. Remember that the wavelength of light varies from about 400 nm (for violet light) to 700 nm (for red light); the diffraction limit was not a problem until the mid-eighties, when the typical feature size first began to drop below one micron. The easiest way to push resolution down was to use shorter-wavelength light sources such as mercury discharge lamps operating in the near ultraviolet. More recently, excimer lasers provide intense sources of radiation deeper in the ultraviolet. Another more subtle scheme to improve the resolution is to use a phase-shifting mask, which, by modifying the thickness of material the light has to traverse on neighbouring parts of the mask, exploits interference effects to increase the effective resolution. It is a combination of laser ultraviolet light sources and resolution enhancement by phase-shifting masks that has made possible linewidths less than 100 nm.

The final stage in the transfer of the pattern from the designer's computer to the physical design of the integrated circuit is etching. For all the

high-technology trappings of a semiconductor fabrication plant, this process
would be completely familiar to an eighteenth-century engraver (see figure 3.2).
The wafer is exposed to a corrosive liquid, which eats away the parts of the wafer
that are exposed, but leaves unaffected those parts that are covered by a resist.

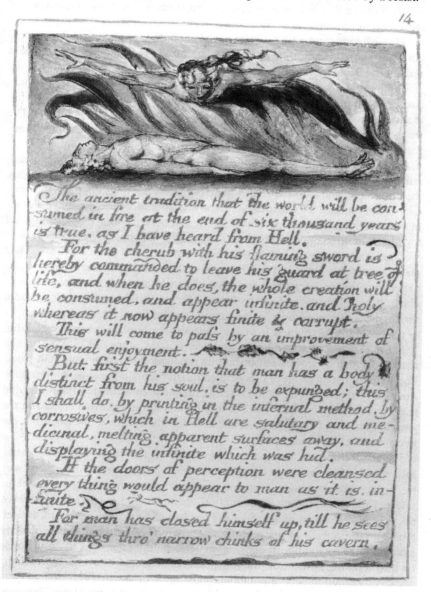

Fig. 3.2 A plate from William Blake's illuminated book *The marriage of Heaven and
Hell.* The plates were made by the direct writing of a resist on copper, followed by etch-
ing with acid. Reproduction by permission of the Syndics of the Fitzwilliam Museum,
Cambridge.

Different materials have different etching solutions. One of the most useful is hydrofluoric acid, which has the property of dissolving silicon dioxide. This property has been used for many years to make designs in glass; no Victorian pub in Britain is complete without floridly-decorated etched windows, the words 'Lounge Bar' being emphasised by garlands and curlicues picked out by the action of hydrofluoric acid on glass.

Hydrofluoric acid is not a strong acid according to chemists' definitions, but its property of corroding the usually inert materials glass and silica give it a daunting reputation. This reputation is augmented by the fact that it is, among inorganic acids, uniquely toxic and damaging in its effects on human tissues. Laboratory safety courses always include gruesome pictures showing the way in which hydrofluoric acid burns deep down into tissue and bone. The major effect these courses have had on me has been that, whenever I have handled the stuff (dressed, of course, in full protective clothing, including full face mask, gloves and apron), I can never sleep the following night for psychosomatic itches up and down my arms and hands. This should remind us that, despite the ultra-clean image of the semiconductor industry, it does get through rather a lot of very hazardous chemicals. The solvents used for cleaning and developing photoresists, the strong acids and acid cocktails used for etching, all together add up to quite a substantial environmental burden.

There are, however, ways of etching that do not require the use of wet chemicals. Aside from any environmental advantages, these actually work better for creating very small structures. The trouble with wet etches is that, although a resist may protect the layer underneath it from direct attack from above, the acid may creep in round the side and undercut the resist at the edges of the pattern. The solution is to use some variety of plasma-assisted etching. A plasma is formed when a very big electric field is applied to a gas; if the field is big enough then the electrons are stripped off from the gas molecules. The resulting ionised gas is both electrically conducting and, quite often, chemically reactive. The process is essentially a controlled version of what happens in a lightning bolt, when a very high voltage between a cloud and the ground ionises the air molecules, allowing the voltage to be discharged to the Earth's surface.

To these three basic operations—laying down thin films, patterning them, and etching them, we need to add one more. To get the electrical properties of the semiconducting layers right, we need to infuse very small amounts of selected impurity atoms—dopants—into them. There are two ways of doing this. The simple one is to expose the layers to a low-pressure gas of the desired impurity; the surface of the semiconductor will take up some of the impurity atoms if it is heated up to 1000 °C or so. How far the atoms can move depends on how long the semiconductor is exposed to them and how high the temperature is, but we are generally talking about an hour or so at elevated temperatures for the impurity atoms to move a micron or so into the layer. The other way of introducing impurity atoms is much faster and more precise. In ion implantation, the atoms are ionised, speeded up in a small particle accelerator and physically hurled at the surface of the semiconductor. A typical speed might be half a million metres per second—at this speed the ion will penetrate tens or hundreds of nanometres into the material before it is brought to a halt by its frequent collisions with the atoms of the semiconductor.

The advantage of ion implantation is that it is a fast and precise way of putting atoms where you want them. The disadvantage is that, in the process of hurling these energetic ions at the surface of your semiconductor, you inevitably do some damage. Atoms are kicked out of their proper positions in the crystal lattice, and these defects will severely impede the flow of electrons through the material. The damage needs to be repaired by annealing—raising the material to a high enough temperature that the atoms can move around and find their proper places again.

These are the basic processes that are needed to make an integrated circuit—depositing thin films, patterning the layers, etching away selected parts, and adding impurity atoms. It is the repeated combination of many successive simple steps that goes to make a complicated integrated circuit like the microprocessors and memory chips in a personal computer.

Moore's law and beyond

Since the introduction of integrated circuits, the smallest size of a feature has decreased at a steady rate of 13% per year. This year-on-year trend to miniaturisation has led to the vast increases in computer power that we are all familiar with, and to the rapid obsolescence of computing equipment. Memory is bigger and cheaper every year; the unit cost of computer memory has recently been halving every two years. As feature sizes get smaller, computers get more powerful for two reasons: ever more components can be squeezed onto a single chip, and as the components get smaller they work faster. As a result of this, computer power has doubled every eighteen months. This technology trend of exponential growth in computer power was first noticed by Gordon Moore, the founder of the semiconductor giant Intel, and it is often referred to as Moore's law.

Moore's law is not, of course, a law at all, in either the scientific or legal senses. It is instead a testament to the power and efficacy of engineering to make continual incremental improvements to a well-established technology. Successive obstacles to further miniaturisation have been overcome by a series of ingenious solutions. A major barrier has been in photolithography. The wave nature of light leads to diffraction effects which limit the precision with which lines can be defined; we have seen that the development of successively-shorter-wavelength ultraviolet light sources and the invention of the phase-shifting mask have helped to push minimum dimensions to smaller and smaller sizes.

Are there any ultimate limits to how small patterning technology can go? Continuing the trend of reducing the wavelength of the radiation used for the patterning, we can think of lithography as using very short-wavelength ultra-violet rays or X-rays. The problems in doing this are that it is much more difficult to find sources of this kind of radiation, and it is even more difficult to focus and reflect the beams. Nonetheless, none of these problems are likely to be insuperable given enough money and engineering effort.

X-ray lithography has many attractive features. Using radiation with a wavelength of around 1 nm means that diffraction will cause no problems. Resist materials that are made more or less soluble when exposed to X-rays are easily found. The big difficulty is finding a source of the X-rays. The X-ray machines used at the dentist or in hospitals are much too weak to be useful, and lasers that emit X-rays have yet to be discovered. The most likely source is from electron synchrotrons. These are the same bright sources of X-rays that are nowadays used to study the structure of molecules by X-ray diffraction. An electron synchrotron consists of a ring in which electrons are accelerated up to very high speeds; as a result of this acceleration they emit electromagnetic radiation. The usefulness of synchrotron radiation stems from a feature of Einstein's special theory of relativity. From the point of view of the electron, it emits radiation uniformly in all directions. But from our point of the view, because the electron is going at a speed which is very close to the speed of light, space and time are distorted so much that all the radiation is emitted from the electron in two narrow, intense beams.

These intense X-ray beams, which include a wide range of wavelengths, are ideally suited for X-ray lithography. The only trouble is that a synchrotron is a major piece of engineering, which is likely to be prohibitively costly, even by the standards of the electronics industry. The synchrotrons that are used for doing X-ray diffraction are often run as national or international facilities. Can they be miniaturised and reduced to a price that will make them economically viable? This remains to be seen.

Another way of beating the diffraction limit whose usefulness we saw when discussing microscopy is to use, not electromagnetic radiation, but electrons. In a scanning electron microscope a beam of electrons is focused to a point and moved across a surface. If that surface undergoes a chemical change when it is exposed to the electrons, then we can write a pattern with the electron beams and use that as a basis for lithography. This technique is known as electron beam lithography. Electrons at the energies used in electron microscopes have a very small wavelength, so diffraction is not a problem—in fact, the limitation on resolution arises from the properties of the resists that are used.

Electron beam lithography is in fact a fairly well-established technique, and for somewhere between half a million and a million pounds one can buy a fully-functional electron beam system. One major use of the technique is to make the masks that are used for the photolithographic process. The circuits are designed on a computer; the computer-aided design package used for doing this produces designs for all of the masks needed for each photo-lithographic step, and a computer controlling the electron beam lithography system writes the pattern directly onto the masks. Another use for electron beam lithography is in the laboratory, where prototype designs for ultra-small devices can be fabricated.

Electron beam lithography is very well suited for producing masks—the originals from which many copies are made by photolithography—and ideal for making one-offs for the physics laboratory. What it is less good at is making production runs of large numbers of complicated integrated circuits. The key difference between electron beam lithography and photolithography is that, in electron beam lithography, the circuit is built up line by line, one component at a time, but in photolithography the entire circuit, with its millions of components, is created at once. This difference is summarised by saying that electron beam lithography is a serial process, while photolithography is a parallel one. This distinction between serial processes and parallel ones is a crucial one. It is obvious that moving a technology from the laboratory into volume production is going to be much easier for parallel processes than serial ones, so making this distinction is going to be vital when one is evaluating the potential of new nanotechnologies.

Direct writing

Is there an alternative to lithography, in which we make nanoscale structures directly by manipulating the atoms? Lithography involves a complex sequence of processes—the selective exposure of the wafer, the development of the resist, and the etching away of the exposed pattern. Could we not operate in a way more directly analogous to the drills, lathes, and milling machines of a macroscopic workshop, directly forming our shape from a block of material?

One way of doing something like this is provided by focused ion beams. We have seen that ions—the charged version of an atom that is formed when one or more of its electrons is removed—can be accelerated in an electric field and fired at the surface of a wafer. If the number of ions is relatively low, then the ions are implanted near the surface of the material. But if we have a beam of very many energetic ions focused to a point, then they can physically drill away their target material. Micro-machining using focused ion beams has proved its worth in making improved tips for scanning probe microscopes, and in repairing damaged photolithography masks.

But the most spectacular demonstrations of atom-by-atom manipulation have been made using scanning tunnelling microscopes (STM). In Chapter 2 we discussed how the interactions between the tip of a scanning probe microscope and a surface can be large—large enough to dislodge an atom or molecule from the surface, allowing one to pick it up and deposit it somewhere else. In this way the ultimate nanostructures can be made—structures which are defined and laid down atom by atom. But it is this precision that also makes the technique impractical for manufacturing—because one can only move one thing at a time, the time it takes to make a structure is very long. Atomic manipulation with an STM is the ultimate in serial, as opposed to parallel, processes, and it is difficult to see how this could be different.

Cheaper, smaller, more curved—soft lithography

Every year computers get faster, but they do not seem to get much cheaper. The corollary of Moore's law, which states that computer power grows exponentially with time, seems to be that the cost of building a semiconductor fabrication plant grows exponentially with time as well. So it may be that the continual advance of CMOS technology to smaller and smaller sizes will be halted, not so much by the inability to overcome physical laws, but the even bigger hurdles that economics places in the way of progress.

The economics of the semiconductor industry are brutal; it has effectively become a prisoner of its own success. The capital cost of a semiconductor fabrication plant is huge, but the rate of progress in the industry is such that, even as the plant is being designed, its owners already know that it is doomed to obsolescence in just a few years. This gives a very narrow time window in which a return has to be made on the investment. The high capital cost and short working lifetime of plants leads to over-centralisation of production and instabilities of supply. This makes the economics of the business even more unpredictable as the prices of commodity integrated circuits, like dynamic random access memory chips, tend to fluctuate wildly.

One of the major reasons for the high cost of semiconductor plants is that current processes demand perfection, and perfection does not come cheaply in any walk of life. For an integrated circuit to work at all, every component in it must work as expected. A single short circuit makes the whole chip valueless, but when the components of the circuit are only 100 nm wide the tiniest speck of dust can cause a failure. So chip manufacturing needs to be carried out in clean rooms in which the air is most rigorously filtered. Every material used in the process needs to be ultra-pure. The smaller the dimensions of the device to be made, the more stringent the requirements for purity and cleanliness become, and the more expensive it becomes to meet these requirements.

Are there any ways of escaping this cycle? Two developments look particularly interesting. The first is probably a prerequisite for the second; it is the introduction of new architectures for chips that can tolerate a much higher level of defects than current devices. The idea here, which has been developed most fully by James Heath at UCLA in collaboration with Hewlett-Packard, is to design a hierarchy of sub-systems. At each level in the hierarchy, the circuit would test all of the sub-systems and rewire itself to exclude the sub-systems that proved to be faulty.

The second development moves away from the expensive photolithographic processes and fabrication plants to some much more cheap and cheerful ways of patterning surfaces. These methods are collectively known as 'soft lithography', and have been extensively developed by George Whitesides at Harvard University.

If photolithography harks back to the early days of photography, soft lithography recalls simple printing methods. Most offices have a selection of rubber stamps, with which one can mark envelopes 'urgent' or 'confidential'. These stamps work well because rubber is soft enough to adapt itself to the surface that one is stamping, but mechanically tough and resilient enough to sustain the small relief features that are needed for the letters to be reproduced well. Soft lithography relies on rubber stamps that are no different in principle to these office stamps. For the 'ink', one needs to use some kind of material that forms a layer one molecule thick when it comes into contact with the surface to be patterned. Whitesides has found that the best inks are a class of molecules known as 'thiols', and the best surfaces are coated with gold. Thiols are organic molecules, about the size of a soap molecule, with a group consisting of a sulphur atom and a hydrogen atom at one end. (The most common thiol is, unfortunately, the compound mercaptan, which is most renowned for its revolting smell of rotting cabbages.) When a thiol group comes into contact with gold, it strongly bonds to it; so where the thiol ink on the relief parts of the stamp comes into contact with the surface, a single molecular layer coats the gold, faithfully reproducing the pattern on the stamp.

The advantages of soft-lithography methods are that they are very straightforward and cheap to implement, yet they are capable of reproducing features on surfaces with a resolution of less than 100 nm. They are very flexible in the sense that many different compounds can be used as inks, allowing one, for example, to obtain chemical patterns, where different regions have different reactive chemical groups at the surface, as well as relief patterns. They are also literally flexible; because the stamp is made from rubber, patterns can be transferred to curved surfaces. These advantages mean that soft lithography has proved very popular in the laboratory, where it has allowed sub-micron patterning techniques to be used in fields of science other than electronics, like cell culture and tissue engineering, where conventional photolithography would be much too expensive. Soft lithography has not yet found large-scale use in manufacturing, though, and so it remains to be seen how important it will prove in the world of engineered products.

Making things besides chips—MEMS and NEMS

The planar manufacturing processes that have been developed for making integrated circuits are brilliantly efficient at turning out vast numbers of identical devices with very small components. It is natural to ask whether one can use these techniques to manufacture other devices and machines besides electronic circuits. The answer is that one can, and many such micro-electro-mechanical systems (MEMS, to the initiated) devices have been demonstrated and commercialised.

Some of the best known of these systems are the arrays of individually steerable mirrors that have been used as optical switches in optical communication systems, while most car airbags are set off by MEMS acceleration sensors. In these commercial systems, feature sizes tend to be in the range of tens to hundreds of microns. As we have seen, silicon micro-electronics has now entered the regime of nanotechnology, with minimum feature sizes falling below 100 nm. This technology is based on operations that are carried out on thin films on flat substrates—planar manufacturing. We can use just the same sort of processes to fabricate other types of nanoscale devices besides electronic circuits on the nanoscale. What could we do with such nano-electro-mechanical systems (NEMS)?

One of the most prominent developers of NEMS has been Harold Craighead at Cornell University. His laboratory has been responsible for some virtuoso displays of technique, most spectacularly making 'the world's smallest guitar' (see figure 3.3). This was made from silicon, with a total length of 10 microns. Each of the six strings was 50 nanometres wide; they do resonate, like a full-size guitar, but because of their tiny dimensions the pitch of their resonant frequencies is much higher than audible sound, being more comparable to radio waves.

The world's smallest guitar was a *jeu d'esprit* rather than a serious bit of science, but there are important potential applications of such tiny, free-standing mechanical parts. They can be used for the very sensitive measurement of small forces; a NEMS resonator is easily able to weigh a single biological cell (strictly, it detects its mass by measuring a change in the resonant frequency). Nanoscale channels and pipes can be used to sort and analyse chemicals at the

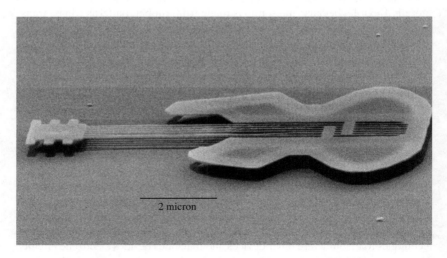

Fig. 3.3 'The world's smallest guitar'. Made by electron beam lithography from silicon, each string is about 50 nm wide. Photograph by D. Carr and H. Craighead, Cornell University, courtesy of H. Craighead.

level of single molecules—an area known as *nanofluidics*, by analogy with the existing techniques of microfluidics.

Are these NEMS techniques going to be the way in which functional nanoscale machines are made in the future? It is certainly likely that these fabrication methods will move out from the electronics industry, where they have been so effectively perfected, into other areas like nanofluidics. But what the world's smallest guitar reminds us is that size is important in determining the way things work. The guitar cannot produce musical notes, because it is too small. Fluid flow through the channels of a nanofluidic device is going to be very different in character to the plumbing in our houses. As things get smaller, their surfaces, and the way they stick to each other, become much more important. These differences in the way things behave on very small scales are what we turn to in the next chapter.

4
The Brownian universe: physics at the nanoscale

Introduction

People who write about the history of ideas know that great works of literature can change the way people think. But popular films, trashy genre novels, and cheesy television shows can change the intellectual climate just as much as high art can. I am sure that the way people think about nanotechnology now has been profoundly changed by the 1966 science fiction film 'Fantastic voyage', later made into a novel by Isaac Asimov and spun out into a Saturday morning television series. The heroes of the film have to save the life of a scientist who knows vital secrets, but who lies in a coma following an attack by the bad guys. He has a blood clot on his brain, inoperable unless the brain surgeon and his team are shrunk to the size of microbes and injected into the scientist's bloodstream in a similarly sunken submarine. After a series of misadventures, our plucky team manage to destroy the blood clot with (shrunken) laser guns and swim to safety, after their submarine is consumed by a white blood cell.

Almost everything about this scenario is quite preposterous, but in quite an instructive way. There are certainly some good things to be learnt from it. There is a real intuitive sense of the sheer amount of space in the nanoworld, just how much room there is at the bottom, to adapt Feynman's words. There is also a compelling visual picture of the complexity of biology at the cell level (even though the primitive state of special effects when the film was made does mean that a lot of the time the tiny submarine seems to be sailing inside a giant lava lamp). What is wrong with the scenario is that it is completely unphysical—if we did manage to shrink a submarine in this way then it just would not behave in the way in which it is shown. But, as usual, it is better to be interestingly wrong than boringly right, and we can learn a lot about the physics of the nanoworld by asking ourselves what is wrong with the 'Fantastic voyage' scenario.

So let us imagine that we really could take a little submarine and shrink it down to cellular dimensions, and then inject it into the bloodstream of our patient. What would our intrepid nanonauts encounter when they turn the engine on and prepare to explore their new nanoworld?

The first thing that they would notice is that they would grind to a sticky halt, and no matter how much they turned up the engine power they would not make much in the way of forward progress. Water, at this scale, is not the freely-flowing fluid we are used to at human dimensions; instead, they will find that their environment is a sticky morass, many times thicker than syrup or treacle.

But just because our submarine cannot go forward, does not mean that it is stationary. Our nanonauts will find their vessel pitching and rolling, jerked here and there by jolts of all kinds, seemingly coming at them out of the blue.

And what is that on the windshield? It is like driving through a swarm of midges, the squashed corpses of the unfortunate insects accumulating on the screen. The nanonauts' formerly clear view is blurred by little blobs of sticky stuff, and no matter how hard they scrub with the windscreen wipers it just will not come off. Soon, all the surfaces of the vessel are coated with a resilient layer. Analysis would reveal that what we have here are protein molecules, which stick to the surface, unfold, and stick together to form a robust two-dimensional film.

Finally, the jerking motion has carried us near a wall. Our steering is useless and we are effectively without power, so we can do nothing as we are drawn inexorably toward it. We come into contact and stick to it, and nothing we seem to be able to do can push us away. Our fantastic voyage has come to an ignominious end.

Physics is different at different length scales. By this I do not mean that the fundamentals of physics are different. Ultimately, all phenomena must be explained by the same sets of equations. What I mean is that the way things behave is different, because different aspects of physics are important. We all have an intuitive notion of physics. It is not that we know the equations; but if we throw a ball, if we make a splash in a bucket of water, if we watch someone fall over, then we know in an intuitive way how things are going to unfold. This is the kind of physics that it is important to get right in putting together a video game; when we cannot get our car round a bend and it flies off we expect it to behave in a realistic way, and the programmers go to some lengths to make sure that this happens. They do not do this by making every atom in the car obey Schrödinger's equation, instead they use a series of approximations that are correct for the scale that they are working at. When we change the scale the necessary approximations have to change too.

Fluid mechanics

I am a very bad swimmer, and it is very obvious to me that to make forward progress through water one has to expend energy. In my case this energy is multiplied by all kinds of unnecessary splashing around, but even the best swimmers, swimming underwater, even the sharks and dolphins that seem so effortless in their motion, must continually use up energy to keep on moving forward at a constant speed. This is different from the motion of bodies in free space—Newton's first law of motion tells us that in an ideal situation a body will keep on moving in a straight line at a constant speed, without any input of energy being needed, as long as no force opposes the motion. So where does the force that opposes the motion of a swimmer come from? There is no doubt that there is such an opposing force; if you stop swimming then you stop moving. It turns out that for motion underwater[3] there are two quite distinct origins of this force. But the relative importance of the two types of impediment to motion depends very strongly on how big you are. This is why, for a bacterium, water feels like a very different kind of stuff to be moving through than for a human swimmer.

The first impediment to motion arises simply from the fact that water has mass, and you have to apply a force to it in order to make it move itself. When you are swimming though water, of course, this is exactly what you have to do. The water ahead of you needs to move out of your way, so you have to apply a force to it to get it to do this. This is why it hurts to jump from a great height into a swimming pool and land on your belly. The water needs to be moved out of your belly's way; to do this your belly needs to apply quite a large force to the water, which is obliged, in turn, to apply the same force back to you. This results in the pain that accompanies the embarrassment of an inexpertly executed dive. It is the inertia of the water—the resistance of any massive object to being put into motion—that impedes motion through it.

The second impediment has a more subtle origin. It is in the property of fluids known as viscosity. This quantity expresses the degree to which, when we try to move one layer of molecules against another layer, we have to use some energy; this energy is converted into heating the fluid. The origin of viscosity at the level of the molecules themselves is actually rather obscure; it is a type of internal friction that requires quite sophisticated theory to understand. But the effects of viscosity are obvious enough. Molasses, treacle, golden syrup, or clear honey all have a very high viscosity, and this is obvious because inserting a spoon into the syrup pot is quite difficult, and having taken out a spoonful, when we try and pour the syrup off the spoon it flows off very slowly. In fact, there is a convenient set of units in which the viscosity of water is about 1; in these units syrup might have a viscosity of tens or hundreds of thousands.

It is obvious that going swimming in a vat of treacle would be a very different experience to swimming in water. It would be very much harder work, for a start, but there are more fundamental differences, too; swimming strokes that work well enough for water simply would not produce any forward motion at all in treacle. The reason for this is that in water the major impediment to forward motion is the water's inertia, while in treacle it is the viscosity that slows one down. The strategies you need to employ to overcome fluid

---

[3] That is, where you do not have the complication of making waves, which substantially slow down both swimmers and ships.

inertia are quite different to the ones that work for overcoming viscosity. So if we were making a submarine that operated in treacle then we would have to use a completely different design for the propulsion system than we would for water.

What is much less obvious, but of central importance for nanotechnology, is that the properties of any fluid, including water, become increasingly dominated by viscosity as the size of the object moving through it becomes smaller. If we were shrunk by a factor of 1000 or 10 000, then water would no longer feel like the free-flowing fluid we are familiar with, but instead it would take on the consistency of treacle.

To see that this is so, we need to be more precise about the way in which the force resisting the motion of an object through a fluid depends on the size and speed of the object and the density and viscosity of the fluid. The inertial force arises from the rate at which we have to give the fluid momentum. The volume of fluid we have to move out of the way in a given time interval varies like the velocity times the area of the object, and its mass is its volume times its density. Its momentum is the mass times the velocity, so the rate of change of momentum is proportional to the density, the size squared, and the velocity squared. In symbols, we can write this as $F_{inertial} \propto \rho \, a^2 \, v^2$, where our object's size is a, its velocity is v, and the fluid's density is ρ.

The force due to viscosity depends, not on the mass of fluid we have to move away, but on the rate at which we are making adjacent layers of fluid move past each other. In fact, the viscous force per unit area is simply proportional to the viscosity and to the rate at which the velocity of the fluid changes with distance. So we have a viscous force proportional to the area of the object, proportional to the velocity divided by the size of the object, and proportional to the viscosity. Again, writing this in symbols, we have $F_{viscous} \propto \eta \, a \, v$ where η is the viscosity of the fluid.

Now we can judge whether the force due to inertia is going to be more important than the force due to viscosity. If we divide the viscous force by the inertial force, then we get a number that will be much smaller than one if it is inertia that dominates. If the number is much bigger than one, on the other hand, then it is the viscosity that will be most important. We could call this the molasses index; the bigger this number, the more treacle-like will the behaviour of the fluid be as we try to move an object through it.[4]

We can now write down how our molasses index varies with the size and velocity of the object that is trying to move, and the viscosity and density of the fluid; it is proportional to the viscosity as one would expect, but it is also inversely proportional to the fluid density, the velocity, and crucially, the size. Imagine a bacterium about a micron big swimming in water. Because it is so small, the molasses index is very large. To get an equivalently big molasses index at the human scale we would need to have a liquid with a very high

---

[4] In fluid mechanics, it is conventional to define a number that is the inverse of this, the Reynolds number. The lower the Reynolds number, the more dominant is the effect of viscosity.

viscosity. A bacterium is roughly one-millionth of the size of a human; to a bacterium, water feels one million times more viscous than it does to us.[5]

What would life be like if we were immersed in a pool of molasses? Extremely inconvenient, is the obvious and entirely accurate answer. The strategies that we use somewhat inexpertly, and creatures like fish and dolphins use much more efficiently, to get around in the water only work when it is inertia that is resisting motion. In a viscosity-dominated situation things are very different. We rely, in swimming, on the fact that we can make a stroke with our arms that gives us a big jolt of forward momentum, and then we can bring our arms back to the starting position in a way that minimises the amount of backwards momentum we get. In a simple sculling stroke, fast motion on the down-stroke gives us a sharp kick forward. If the return stroke is slow, then we do not lose too much ground. In a viscosity-dominated environment you cannot do this. You would sweep your arms back, and move correspondingly forward, but as you brought your arms back to their starting point, no matter how slowly, you would be pushed back to your starting point. A more sophisticated swimming stroke depends on changing the effective cross-section of your arm or leg from the down-stroke to the return stroke, but even this strategy is much less efficient in viscosity-dominated conditions, because the drag force imposed on an object moving in fluid doesn't depend very strongly on its shape. In treacle, even the most expert swimmer would not be able to do much more than expend a lot of energy rocking backwards and forwards around one spot.

How do bacteria and other nanoscale objects get around at all? It turns out that, in a situation dominated by viscosity, the best way to make forward progress is by a twisting motion. If the 'Fantastic voyage' submarine had been equipped with a giant corkscrew attachment at the front, then as the corkscrew twisted it would be able to pull the submarine through the treacly water. It is not a very efficient way of making progress; you need a long corkscrew and you have to put a lot of energy in to make it go round, but it is more or less the best you can do. This is the solution that nature has evolved for bacteria; they are equipped with long, whip-like tails called flagella, which are rotated by tiny, molecular-sized rotary motors, twisting them round through the fluid.

Flying nanobots?

We have talked about a watery environment up to now, but could a nanoscale robot fly? In the novel *Prey*, by Michael Crichton, the climax occurs when our hero is pursued by a self-organised, solar-powered swarm of flying nano-engineered robots, the result of an unwise attempt to fulfil a US Department of Defense contract in a hurry. It is not clear how big they are; presumably less than a hundred microns or so, as they are not discernible to the naked eye as individual particles, but they are definitely bigger than the medical version of the robots, which are 500 nm in diameter (one-tenth of the size of

---

[5] This argument neglects the fact that bacteria will not be swimming as fast as humans, which makes the situation even worse.

a red blood cell). No matter how big they are, they can fly faster than our hero can run. Does this make sense? How small is it possible to make a robot that flies?

To answer this question, it is worth looking to nature. The smallest flying insect seems to be a parasitic wasp called dicopomorpha echmepterygis, which is a little more than a tenth of a millimetre big. Is it that there are no smaller insects because it is physically impossible, or simply because no evolutionary advantage would accrue to a flying creature that was any smaller? The answer seems to be the former, and that this represents pretty much the smallest size object that one could imagine flying in an atmosphere like ours. To understand why, many of the same considerations that underlie the scaling behaviour of swimmers recur, especially the increased importance of viscosity for small objects. But, as Daedalus found out, flying is significantly more difficult than swimming, because in addition to finding the power needed to move forward, a flier needs to generate enough lift to stay in the air.

The physics of flying insects is notoriously difficult to understand, so much so that, on a roughly biennial basis, newspapers announce (to the accompaniment of much sniggering) that scientists have proved that bees cannot fly. What is clear is that the engineering challenge of making an object the size of the average insect (the five or ten millimetre length of a fly or a honeybee) that flies is very much more difficult than making a passenger aeroplane.

A wing generates lift, but it also creates drag; a flying machine needs to supply energy both in order to keep itself aloft and to overcome the drag to keep itself going forward. As wings get smaller, the effect of viscosity gets larger. The molasses index, which we introduced to measure the relative importance of viscosity in flow is around one-millionth for an aeroplane, but for a small insect this falls toward one-hundredth. Viscosity is ten thousand times more of a problem for an insect than for an aeroplane. This is reflected in the best achievable ratio of the lift to the drag force, which is a good measure of how effective a wing is. A reasonable value for lift/drag for an aeroplane wing might be 25, but for a fruit-fly this has fallen to 1.8, the deterioration being almost entirely due to the pernicious effects of viscosity at small sizes.

An insect can make some compensation for its small size by beating its wings faster; while a fruit-fly beats its wings around 200 times per second, a midge can manage about 1000 beats per second, accounting for the high-pitched whine which is often the only warning one has that one is about to be bitten. But the cost in energy consumption of doing this is prodigious.

There are, of course, many tiny biological objects that do choose to travel by air, perhaps the most obvious being the pollen that causes such difficulty to hay-fever sufferers in the summer months. Pollen grains are typically some tens of microns in size, but fungal spores can be quite a lot smaller. Clearly, the adaptive purposes of both pollen and spores are best served the further they can travel, but they do not fly, they drift. Rather than actively seeking to stay aloft, pollen and spores are designed to minimise their sinking speed by increasing the drag force that air imposes on them. In this way they can travel long distances on the slightest of breezes.

Brownian motion

We have seen that, at the nanoscale, water is not the free-flowing fluid we are familiar with at the macroscale. Nanoscale objects in water find themselves in a sticky, treacle-like fluid in which any kind of directed motion is very difficult. Does this mean that the nanoworld is static and without any kind of rapid motion? It emphatically does not; although directed, purposeful motion is difficult, all nanoscale objects undergo a different type of motion which is random rather than directed. If you were to look at any small object then what you would see would be a continuous jiggling motion. This motion is called Brownian motion; it is inescapable at the nanoscale. It is an inevitable consequence of the fact that matter is made up of atoms and molecules.

Suppose that some catastrophe were to destroy all scientists and all of our records of science, but we had the opportunity to choose one single scientific idea to pass on to the survivors. What idea or equation, if passed down in this way, would give our descendants the fastest head start in reconstructing the achievement of today's science? This question was posed by Richard Feynman and his answer was not $E = mc^2$, or the path integral formulation of quantum mechanics, or Newton's laws of motion, but the simple idea that all matter is made up of atoms. This idea is so familiar and so ingrained that we forget how revolutionary and how counter-intuitive it is. Modern technology has given us fairly direct ways of seeing atoms, so it is also easy to forget that, for more than two millennia after the idea was first proposed by the Greek philosophers, any evidence for it was indirect and inferential. Even at the beginning of the twentieth century, serious scientists could still be sceptical about the atomic hypothesis; the phenomenon that finally convinced the doubters was Brownian motion. Brownian motion had been observed in the first half of the nineteenth century, by the botanist Robert Brown. He had been observing plant pollen in his microscope, and was surprised to see the tiny particles undergoing a continual, frantic dance. There was much speculation on the cause of the phenomenon throughout the nineteenth century, but the matter was settled in the first decade of the twentieth century. Jean Perrin carried out painstaking experimental analyses of the phenomenon between 1906 and 1908. These confirmed the theoretical analysis that Einstein had done a year or two before, and provided the final, definitive proof that matter is not continuous, and infinitely subdividable, but instead is made up of discrete units—atoms.

The explanation of Brownian motion, then lies in the idea that matter is made up of atoms, and that these atoms are in a state of constant motion. That kind of energy we know as the heat of some piece of matter is simply the undirected, random energy of motion of the atoms making up that matter. In a liquid or a gas, if a larger particle is suspended amongst the atoms then it will itself dance around randomly. This is because the smaller atoms are constantly colliding with the body. On average, it will receive an equal number of blows coming from the left as from the right. But in any short space of time, the discrete nature of atoms means that there may be a momentary imbalance in the number of blows from each direction. Like a fat man being jostled in a crowd, at one moment there will be a net push to the right, then a little later there will be a little shove to the left. Why does Brownian motion affect small particles more than big ones? There are two reasons. Obviously, the bigger the size disparity between the particle being jostled and the atoms doing the jostling, the bigger the effect of a single collision. In addition to this simple size effect, there is another effect relating to the way the numbers of collisions from each direction fluctuate. If, in a given interval of time, there are only a few collisions, then the relative imbalance between the number of collisions from each direction is likely to be larger. These two factors combine to mean that the smaller the particle, the larger the effect of Brownian motion. Small particles are tossed and buffeted by the ceaseless impact of the jostling atoms, while large particles sail serenely on.

How can we say anything at all about how a pollen particle moves when immersed in water, when we realise that to understand its motion we need to keep track of the random motion of a huge number of water molecules, all of which have impossibly complicated trajectories, involving multiple collisions with each other and with our pollen particle? Conventional mechanics tells us how to deal with the collision of one billiard ball with another, but if we have thousands, hundreds, or even tens of balls then this kind of calculation descends quite literally into chaos. In principle, classical mechanics is deterministic, in the sense that if we know at one given time where all the particles are, together with their initial speeds and directions of motion, then we should be able to find out where they all are at any time in the future. But, in practise, any uncertainty in our original knowledge of the particles' positions and velocities, or any perturbation that arises from some external influence, like another particle coming into our system from the outside and making an extra collision, will be enormously magnified with time until, very shortly, we lose all practical ability to make any predictions about the motions of our particles thereafter. But if we cannot make definite, deterministic predictions, we can make statistical ones, and these are remarkably simple and robust.

One of the simplest of these predictions also turns out to be the most useful in understanding why Brownian motion becomes so important for nanoscale systems. We know that all of the components of any system will all have a certain amount of heat energy; the molecules of a gas or a liquid are constantly moving around and possess kinetic energy by virtue of that motion. If the liquid contains some larger particles, then they too will be moving around by virtue of Brownian motion, so they will also have some complement of kinetic energy. As the molecules collide with each other and with the suspended particle, energy is constantly being passed about. It turns out that the statistics of this process are such that, on average, each particle has a certain, constant ration of energy, which is proportional to the temperature of the system. This theorem is known as the equipartition of energy.

Because the average energy of any particle due to its Brownian motion depends only on temperature, at a given temperature the average velocity of the particle will depend strongly on its size, because the kinetic energy of an object is proportional to its mass. It is easy to calculate that, for a single water molecule at room temperature, the average velocity due to its thermal energy is about 260 metres per second, nearly 600 miles per hour. Beneath the placid surface of a glass of water, molecules are moving around at the speed of a jet aircraft, constantly colliding with each other and with anything else in their path. A particle 100 nm in size—a small bacterium, for example—at an average speed of 50 metres per second, or 120 miles per hour, is going at the speed of a fast car, while a one-micron-sized particle—like one of the fat globules in a glass of milk—will be making a brisk walking pace, at 1.7 metres per second or nearly 4 miles per hour. Small things go fast—but they do not go very far. In the dense crowd of other speeding molecules, they soon hit something else and bounce back the way they came. The scene at the nanoscale is one of ceaseless, energetic, and rather pointless movement.

But just because the Brownian motion of a particle is random and undirected, does not mean that the particle does not get anywhere. The particle moves a little way in a straight line, hits something, it sets out again in a totally different direction until it has its next collision. This sort of motion is known as a random walk. The distinction between this type of motion and normal motion is this. If you carry on in a straight line, then you go a distance which is proportional to the number of the steps. But if, after each step, you change direction, then you go a distance which is proportional not to the number of steps but to the square root of this number. This means that a random walk is a surprisingly effective way of covering a short distance, but is a very bad method for trying to go a long way.

This has very important consequences for the way that matter gets around in small systems compared to large ones. Transport of matter by this random walk process is known as diffusion, and the way diffusional transport depends on size has huge implications for the way evolution has designed organisms. Take, for example, a small molecule like oxygen. The diffusion coefficient of oxygen in water is 18×10^{-6} cm^2 s^{-1}; roughly speaking, the time it takes a molecule to diffuse a given distance is given by the square of the distance divided by the diffusion coefficient. So an oxygen molecule will move 10 nm in a few nanoseconds, it takes milliseconds to go a micron, but for it to move a centimetre by pure diffusion we would have to wait many hours. This is why humans have blood circulation systems and bacteria do not.

Bigger molecules diffuse somewhat more slowly than little ones; the diffusion coefficient for a sugar molecule (sucrose) in water is about 5×10^{-6} cm^2 s^{-1}, while for a macromolecule like a protein the diffusion coefficient is likely to be rather less than 10^{-6} cm^2 s^{-1}.

A bacterium is small enough that diffusion can be relied on entirely to transport molecules inside it. A bacterium does not need a blood supply to move food and oxygen around, as it can rely on the Brownian motion of these molecules to get them to where they need to go. It can even rely on diffusion to bring it its dinner; Brownian motion will bring food molecules faster to the bacterium than it can consume them. We can think of bacteria as being like cows that do not need to move as they eat, because the grass grows as fast as the cows can munch it. Bacteria can move, but they only need to do this to find a greener field.

The continual collisions that a nanoscale object is subjected to can do more than just make it move around. If the object is at all complicated then the collisions will make it do all kinds of internal motions too. Imagine a nanoscopic dumb-bell—two spheres on either end of a bar. The collisions that one sphere is subjected to will be completely independent of the collisions on the other sphere. Each sphere will be pushed this way and that, and as a result the dumb-bell will twist and flex. The more complicated the object—the more internal structure it has—the more different types of motion it can have. Imagine something like a gearbox—a collection of cog-wheels on shafts with mechanisms to move them in and out. Under the bombardment of molecules that causes Brownian motion, the mechanism would be flexing and bending, the cogs would be spinning first one way and then the other, gears that were meant to mesh would become unmeshed, and components that ought to be kept clear would foul each other.

Brownian motion is inescapable in the nanoworld—you cannot engineer around it. It is a property of the medium, not the object, so there is no material that we can find or cunning design that we can devise which will make a nanoscale object immune to it. It does not even depend on the properties of the medium that much. The dumb-bell we introduced above will still flex, whether it is immersed in water or in air, and the amount it flexes will be exactly the same. This is a surprising result; you might think that in a medium like air, where there are many fewer molecules in a given volume than in water, there would be fewer collisions and therefore the Brownian motion would be less intense. It does not work that way because the molecular collisions have two effects; they set the particles in motion, but they also slow them down. The viscosity of fluids, which we have discovered to be dominating on the nanoscale, and which has the effect of making directed motion very difficult, ultimately has the same origin as Brownian motion. If we try to move a particle through a liquid, then there will be, on average, more random collisions on the upstream side of the particle than the downstream, and this difference in the number of collisions will tend to slow the particle down. In a low-density medium, like air, there are fewer collisions to set the particles in motion, but there is less viscosity to slow them down again once they have been set moving. Remarkably, it can be shown that these two effects always cancel each other out. This means that the average degree of flexing of our dumb-bell does not depend on the nature of the medium it is immersed in, or the way in which the dumb-bell interacts with the medium. It only depends on one thing, the temperature, which determines the average kinetic energy of the surrounding molecules.

But wait a minute . . . if Brownian motion is a result of random molecular collisions, but the amount of Brownian flexing that a dumb-bell undergoes depends only on temperature, then what would happen if we remove the medium entirely? What happens if we put our dumb-bells in a vacuum chamber and pump out the air, so there are no more molecules to collide with it? Surely this would put a stop to the random motion? Alas, no; as long as we can define a temperature for our object it will undergo Brownian motion appropriate to that temperature. The immediate physical cause of the Brownian motion will be whatever mechanism communicates the temperature from the object to the environment outside the vacuum; it is the mechanism by which, if we heated up the outside of the vacuum chamber, the object itself would warm up. In the absence of any matter, the way heat energy would be transmitted from the hot walls to the cold dumb-bell would be electromagnetic radiation—photons. It would be the random emission and absorption of photons by the dumb-bell that would give the little jolts to the dumb-bell that cause its Brownian motion.

The only way to get rid of the Brownian motion of a small object, then, is to cool it down, and keep it isolated from everything around it. When we say cold, here, we do not just mean that it is going to have to live in the fridge. The average Brownian speed of a particle, or the amount of Brownian flexing in a nanoscale structure, varies as the square root of the absolute temperature. To reduce the amount of Brownian motion by a factor of 10, we have to drop the temperature to about 3 K, or $-270\,°C$. As our aim in making nanoscale machines and devices is to interact with the environment, this is almost always going to be impractical. Brownian motion is a feature of the nanoworld that nanotechnology will just have to learn to live with.

Stickiness

In the nanoworld, things stick to each other. In fact, in the macroworld things stick together too, if we can get them close enough to each other. If you take two pieces of glass which are polished very smooth and are quite clean, and bring them together, then they will stick together quite strongly; the reason most surfaces we encounter do not stick together is simply that they are so rough. Even a highly-polished piece of metal will have a surface that looks mountainous on the nanoscale. If we bring two such surfaces together then they only come into true contact in one or two places where the surface protrusions meet. Alternatively, if a material is soft and deformable, it is usually sticky, because when we press it onto a surface it is able to squeeze itself around the protrusions on that surface in a way that gives a substantial fraction of the area to be in close contact. Examples of this include cling-film, which sticks well to almost any clean surface because it is soft enough and thin enough to deform in a way that allows it to follow the contours of the surface it is pressed on. In just the same way, the sticky side of adhesive tape or post-it notes works because the material coated on the film is soft; it conforms to the roughness of whatever you stick it to under the light pressure you apply with your fingers (these materials are known as 'pressure sensitive adhesives'). On the other hand, in an engine or other machinery made of metal, if we put too much load on those parts that move against each other, and let the engine run out of lubricating oil, the parts will stick together irreversibly, or seize up. This happens when the force pressing the parts together is so large that the protrusions that normally keep the surfaces apart are crushed together, and a significant area of the two surfaces comes into true contact. At the nanoscale, these forces of adhesion which we have to work hard to bring into play in the macroworld are dominating in importance. We have to work hard to stop nanoscale objects sticking together, and then staying stuck—irreversible seizing up is going to be a very common fate for poorly-designed nanomachines.

What are the origins of these forces of adhesion? Physics tells us that there are only four fundamental forces. These are the strong nuclear force, the weak nuclear force, electromagnetism, and gravity. Leaving aside phenomena relating to radioactivity and nuclear fission and fusion, the first two forces are not really important in our daily lives (apart from keeping the nuclei of stable atoms stuck together). In our lives, the effect of gravity is all too obvious in keeping us stuck down on the Earth's surface, and keeping the planets and galaxies turning. Almost all of the other phenomena of matter that we see in everyday life—all chemical transformations and all interactions between material things—owe their origins to electromagnetism. In the nanoworld, the dominance of electromagnetism is even more complete. The effect of the gravitational field of the Earth on particles less than a micron or so in size is small, while the force generated by the gravitational interaction of two nanoscale particles themselves is utterly negligible.

The most direct manifestation of the electromagnetic interaction is the attraction between two opposite electric charges, or the repulsion between two like electric charges. This is what gives rise to the classic phenomena of static electricity. Forces between charged bodies can be important at the nanoscale, but there is another type of force, also ultimately electromagnetic in origin, that is completely ubiquitous and arises even between objects that are neutral in charge. Various manifestations of this force are known as van der Waals forces, dispersion forces, or the Casimir effect. Completely negligible at the macroscale, these forces are as ubiquitous and dominating at the nanoscale as gravity is to Earth-bound humans.

To understand the origins of these forces, we need to understand the way in which two apparently quite different concepts dominate modern physics—the idea of the particle and the idea of the field. In Newtonian physics there is a certain primacy about the idea of the particle, with the field needed to explain the way in which forces are transmitted between particles. Maxwell showed that the field could achieve a life of its own, in that vibrations of the electromagnetic field could be essentially self-sustaining. These vibrations are manifest to us as electromagnetic radiations such as radio waves, light, or X-rays, and they have the characteristic that they can carry energy and momentum. In quantum mechanics the distinction between particles and fields is blurred; the energy associated with fields in many circumstances can take only finite values, and these packets of energy can be thought of as particles. For example, the electromagnetic field can sustain waves, and each packet of quantised energy in one of these waves is known as a photon. One of the puzzling and unexpected results of quantum mechanics is that even completely empty space is filled with an infinite number of different electromagnetic waves, each of which carries a definite minimum energy—the zero-point energy. Thinking about this in terms of particles, one has to imagine space as being filled by photons which come into existence out of nothing, and shortly return to the void, returning their borrowed energy. The fact that nature seems to be rather a sloppy accountant of energy, allowing this kind of short-term loan, is expressed in Heisenberg's famous uncertainty relations. The consequence of these is that there is an infinite amount of energy contained even in the empty space of the vacuum.

The existence of this vacuum energy seems paradoxical, even nonsensical. But it has very pronounced and readily apparent consequences. Imagine that we have two parallel metal plates, separated by a few tens of nanometres. If the metal is a perfect conductor of electricity, then no electromagnetic waves can penetrate inside it (this is why a mobile phone will not work if you wrap it in aluminium foil, or for that matter why we cannot see through steel). Although it is possible to sustain electromagnetic waves in the gap between two plates, these waves are profoundly affected by the presence of the metal plates. A wave is completely reflected by a metallic plate; the result of multiple reflections of a wave between two plates is to set up what is known as a standing wave. This is just like the wave that you get on a vibrating string held between two fixed points, such as a violin string. A standing wave like this can only have certain well-defined frequencies; a violin string when it is made to vibrate sounds a note which is made up of a fundamental tone and a set of harmonics. The fundamental tone is given by the condition that half a wavelength fits into the separation between the plates, and the harmonics are waves of higher frequencies in which a whole number, greater than one, of half-wavelengths fit between the plates.

The effect of this is that, between two metallic plates, there are fewer possible types of wave that can be sustained; in free space, we can have vibrations of any frequency at all, but between the plates we are restricted to those frequencies that are harmonics of the fundamental frequency, which itself is determined by the separation between the plates. The reason this is important is that quantum mechanics says that each of these types of vibration has a certain minimum energy. The total zero-point energy we get by simply adding up the number of different vibrations, and saying that each of those vibrations has one unit of zero-point energy. If there are fewer possible vibrations between the plates than in empty space, then the energy associated with the space is decreased when the plates are inserted. If the energy is decreased when the plates are in place, then there must be a force that draws the plates together, just as a force pulls you down to the lower-energy state at the bottom of a hill. The presence of this attractive force is known as the Casimir effect, and we can calculate the value of the force exactly. In fact, the force per unit area is $1.3/h^4$ GPa, where h is the separation of the plates in nanometres. So at one micron separation we have a pressure of about one-hundred-millionth of an atmosphere. This does not sound very much, but the pressure gets very much larger as the separation gets smaller. As soon as the plates start to drift together, the force rapidly increases; by the time the separation is 10 nm the pressure is equivalent to atmospheric pressure, and if nothing was to keep the plates apart then they would inexorably grind into each other and stick.

There are some very mysterious features in this argument. Firstly, it is easy to see that, in empty space, because there is no restriction on the type of vibration that the electromagnetic field can sustain, there are actually an infinite number of possible vibrations. As each vibration carries a discrete packet of zero-point energy, the total amount of energy contained in these zero-point fluctuations of the electromagnetic field—the vacuum energy—is in fact infinite. Even when we restrict the possible types of vibrations of the field by inserting our parallel metal plates, there are still an infinite number of vibrations, but the difference between the two infinities seems to be finite. This seems very fishy, but clever theoretical physicists assure us that it is true. If we have two metal plates, and we let them approach each other, then we could use the attractive force between them to do work. We could think of this as extracting energy from the vacuum to do something useful with. Since the total energy reserves in the vacuum are in fact infinite, this has suggested to some people that our energy problems can be solved by just siphoning energy out of the quantum vacuum. Unless I am missing something, I cannot see that there is any more chance of this scheme producing limitless energy than there is of any of the other designs of perpetual-motion machines proposed over the last few hundred years being successful. Certainly, if we could attach some generator to the two plates then we could extract energy as the plates are pulled together, but we would have to put the same amount of energy back in to get them apart.

We have talked about metal plates attracting each other across a vacuum as a result of the fluctuating field between them, but exactly the same principle holds between any material bodies interacting through any medium (in the more general case, we talk about dispersion forces or van der Waals forces rather than the Casimir effect). The details of the theory are complicated, but the result is very clear—any pair of bodies made of the same material will always attract each other. The attraction becomes significant as soon as the separation is less than one hundred nanometres or so, and it becomes stronger the closer the bodies approach. It is not so much the properties of the materials themselves that result in this force; it is the nature of the vibrations in the electromagnetic field between them, and this is why the effect is so general.

Imagine two ships, broadside on, in a rough sea; between the two ships there will be a little patch of calmer water. Because the sea on the outside of the pair of ships is rougher than the water between them, the two ships will be pressed together. This is in very close analogy to the origins of the dispersion force and the Casimir effect.

If it is a general law of nature that small objects will tend to stick quite strongly together as a result of the dispersion force, then does that mean that nanotechnology is completely doomed as an idea, and that any nano-devices we successfully manage to make will rapidly seize up? No, because there are some strategies we can employ to overcome the dispersion force and keep the components of our nano-devices apart. The physics of these strategies can be quite subtle, but that does not mean that they are not well known already. It is possible to make suspensions in water of very small particles which do not stick together; these are called colloids, and many everyday products employ these tricks to keep them apart.

The first people to master the art of making nanoscale particles and keeping them apart were the ancient Egyptians. They invented what we know now as Indian ink, which is made from carbon black—soot—and water. If you try and make ink by simply mixing soot and water, you will get a useless mess; the tiny soot particles simply stick together and fail to disperse in the water. If you try and write with this mixture, your pen will leave a trail consisting either of slightly dirty water, with not enough carbon in to make it black, or black soggy lumps that are too thick to flow. To make Indian ink you need to add a third component, gum arabic, a resin-like material obtained from acacia bushes. With the addition of this gum, the carbon black can be successfully dispersed in the water to yield an ink, with enough carbon in to leave a strong black line when the ink dries, while staying runny enough to be able to write with.

It took a few thousand years to work out why this ancient method of making ink works. Gum arabic is a natural polymer, a long, chain-like molecule that is soluble in water. What happens when you add soot to a solution of gum arabic is that polymer molecules partially stick to each, nanoscale-sized soot particle. If we were to look at the surface of each particle, then we would see our string-like polymer molecule stuck on at a few points, leaving great loops of string projecting out into the solution. As two soot particles approach, they are attracted to each other by the dispersion force; without the polymer they would be pulled together until their surfaces came into contact, at which point they would be irreversibly stuck. But with the polymer stuck to the surfaces, the great loops of polymer sticking out from the surfaces get in the way, and keep the two surfaces apart.

We can think of this as being akin to the way in which a piece of sticky tape, after it has been stuck to a piece of cloth, is no longer very sticky, because it has been coated with pieces of fluff. Two fluff-coated pieces of sticky tape will stick to each other very much less than they do when they are clean, and the reason for this is obvious; the fluff, itself firmly stuck to the tape, simply stops the two sticky surfaces from approaching close enough to stick.

The mechanism by which fluff stops sticky tape from sticking is, however, a little different to the way polymer molecules work. It is the actual physical rigidity of the tiny fibres of fluff that keeps the surfaces apart, the fibres behaving as springy buffers. But polymer molecules are not rigid at all. A loop of polymer molecules is like a long loop of string; completely flexible and utterly unlike a springy piece of wire. Rather than it being the intrinsic rigidity and springiness of the polymer molecule that keeps the surfaces apart, it is that all-important feature of the nanoworld, Brownian motion, that is at work here.

Imagine two nanoscopic particles, beginning to approach each other under the influence of their attractive dispersion forces. Attached to their surfaces are loops of flexible polymer. Every segment of each polymer chain will be jiggling this way and that, under the influence of the constant bombardment of the surrounding water molecules. The effect of this is that the loops and tails of each polymer molecule will be constantly writhing and gyring under the influence of this continuous bombardment. Sometimes chance will have it that a tail is extended out a long way into the solution, in which case the collisions will tend to cause the molecule to bend and fold up more. At other times a loop will, by chance, be particularly compactly folded near the surface of the particle, in which case the random collisions of the water molecules will tend to unfold it and extend it away from the surface. Left to themselves, the loops and tails of each polymer will be continually flexing and contracting, in such a way that there is some average thickness of polymer around the surface of each particle. Now, if the particles start to come closer to each other than this average thickness, then, when a set of collisions causes a tail or a loop of one polymer to extend, it will collide with the other particle, and tend to push it away. It is the random Brownian motion of the water molecules that is being directed by the polymers to push the two particles apart, keeping them unstuck, in spite of the attractive dispersion forces drawing them together.

This method of stopping small particles sticking to one another by coating them with polymer molecules is known as steric stabilisation, and is common in both nature and technology. Emulsion paint operates on this principle; the main film-forming ingredient is a dispersion of sub-micron polymer particles that are decorated with a coating of water-soluble polymer chains. When the paint is wet, the protruding chains keep the polymer particles apart, but when the paint dries the particles merge to form a continuous plastic film, binding the pigment particles together and keeping the whole paint layer stuck to the wall.

There is one other important way of stopping nanoscale particles from sticking together. This uses the simple fact that objects with the same electrical charge repel each other. If you stroke a cat with a nylon cloth, then after a while the creature's hair starts to stand on end. This is because, by the simple act of rubbing, it is possible to remove electrons from the fur, leaving each hair with a net positive charge. This means that neighbouring hairs then repel each other. So if we can treat our nanoscale particles in some way that means that they all have the same electrical charge, then the repulsion of the different charges will overcome the attractive dispersion forces and keep the particles apart. A common way of doing this would be to add a charged surfactant—a soap or detergent, in other words.

As usual, there are some interesting complications that come about when we operate at the nanoscale, particularly in a watery environment. Water is very rarely pure; it usually contains dissolved salts such as common salt, sodium chloride. These salts break up in solution; each sodium atom gives up one electron to a chlorine atom, and the result is that in solution there are freely-dispersed charged particles, or ions, or sodium and chlorine. The presence of free charged ions in solution means that the surface of any particle, if it is charged, will attract oppositely-charged ions to it, effectively neutralising its charge. Does this mean that this kind of electrostatic force cannot operate in water? Not quite, thanks yet again to the importance of Brownian motion. Although the surface attracts the ions toward it, they are prevented from sticking to it by Brownian motion. Around a charged surface, there is a diffuse cloud of oppositely-charged ions, each of which is constantly being shaken about by Brownian motion, sometimes escaping from the surface, to be replaced by another ion moving in the opposite direction. The thickness of this cloud of ions depends on the concentration of the solution. For pure water, in which the only ions are those that come from water molecules themselves, the ion cloud thickness is 960 nm, meaning that the electrostatic interaction operates at the nanometre scale in a similar way to in a vacuum or air.[6] But at the salt concentration that occurs in our bodily fluids, the thickness of the cloud of ions surrounding a charged surface is 0.8 nm. This means that, at distances greater than this, the repulsive effect of the opposite charges is completely neutralised. Only when the surfaces get closer to each other than this do the charges have any effect in producing a repulsion.[7]

So we can use the repulsion of like electric charges to keep nanoscale particles and components apart, but the efficiency of this mechanism depends very strongly on whether we are in salty water. In fact, we can often take a suspension of nanoparticles that are stabilised against sticking together by charges, and make them unstable by adding extra salt. This is an important process in a number of technologies, where it is known as 'salting out'.

Some final forces that are worth mentioning arise from the special properties of water. Everyone knows that oil and water do not mix. If you mix up vinegar and olive oil to make a salad dressing, then no matter how vigorously you shake the mixture, as soon as you stop, the little droplets of oil start to coalesce to make bigger ones. The bigger ones in turn coalesce until, quite soon, you have a single layer of oil sitting on top of the vinegar. What is the force that causes the oil droplets to combine?

The standard physical chemistry explanation is that oil is hydrophobic—it 'fears' water—and the hydrophobic oil drops band together to keep away from the terrifying liquid. This, of course, is no explanation at all, but what is

---

[6] Though, as the dielectric constant of water is 78.5 the magnitude of the interaction is reduced by this factor.

[7] In fact, at moderate separations it is the Brownian motion of the ions in the neutralising cloud that leads to a repulsion between the plates rather than the direct interactions of the charges.

truly going on is still rather obscure. What can be said is that the unfavourable interaction between water and typical hydrophobic molecules like wax and oil is related, not to the energy of their interaction and the forces between them, but to the state of order imposed on the water molecules by the presence of these foreign molecules. This, in turn, is related to the curious structure of liquid water.

The best way to think about most simple liquids is as being composed of hard, smooth balls. The balls attract each other in a fairly non-specific way, so that they are mostly touching, but being hard, the distance between the centres of pairs of balls is never less than the balls' diameter. Nonetheless, Brownian motion causes them to be in a state of constant motion, stopping them from falling into a regular packing of the kind you get if you stack oranges. What determines the structure of the liquid—how far apart the molecules are on average—is the hard repulsion between two balls when they come into contact.

Water behaves very differently to this. A water molecule consists of a central oxygen atom, bonded to two hydrogen atoms. But the two oxygen–hydrogen bonds are not arranged in a straight line; instead they form an angle of 104.5°. Each of the hydrogen atoms can form a special type of bond with an oxygen atom of a neighbouring water molecule. This bond—known to chemists as a hydrogen bond—is not as strong as a conventional chemical bond, but it has a degree of stickiness that is sufficient to resist being shaken off by Brownian motion most of the time. Each oxygen molecule can form two such hydrogen bonds; one can think of an oxygen atom in water as being in the centre of a tetrahedron, with a hydrogen atom at each corner. Two of these hydrogen atoms are joined to the oxygen atom by conventional chemical bonds, to form the water molecule. The other two hydrogen atoms are joined by the weaker hydrogen bonds, and more properly belong to neighbouring water molecules.

The writer Philip Ball uses a beautifully simple image to help visualise this. Imagine that you are a water molecule, and your two hands are the hydrogen atoms. Your ankles are the two sites on the oxygen atom that can make a hydrogen bond to a neighbouring hydrogen atom. You get the geometry about right by standing with your feet apart, splayed out by a bit more than 90°, twisting your trunk by 90°, and stretching your arms out. Now the rule that describes how water molecules interact with each other is simple; one hand grabs the ankle of one of its neighbours.

Solid water—ice—can be visualised as a kind of human pyramid, in which each molecule is linked to four of its neighbours by these hydrogen bonds, forming a completely regular lattice. When ice melts to make liquid water, not all of the hydrogen bonds break. Most water molecules are still linked to four neighbours, but under the influence of Brownian motion, hydrogen bonds are continually being broken and reformed. A hand lets go of an ankle, and then, moments later, grabs it again, or grabs another free ankle that happens to be in range. It is this dynamism that makes water liquid, and destroys the long-ranged order of the ice.

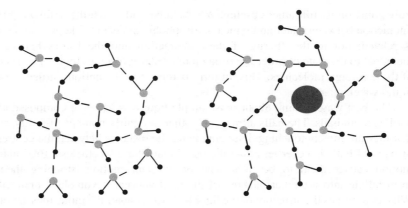

Fig. 4.1 The origin of the hydrophobic interaction. Water forms a continuously changing, disordered network of hydrogen bonds (left). If a molecule is introduced that cannot form hydrogen bonds (right), then the nearby water molecules are forced into a more-ordered arrangement.

Now suppose that we introduce a hydrophobic particle into the water (see figure 4.1). In our picture of hands and feet, it is as if we introduce a giant ball with no hands to grab and nothing to grab onto. In order to find anything to grab onto, water molecules nearby need to turn away from the ball. This limits the freedom which the molecules have to move, forcing them into a more-ordered arrangement than they adopt in the liquid. It is this increase in local order—in terms that will become more familiar later in the book, a decrease in local entropy—that leads to the unfavourable interaction between a hydrophobic particle and water, and the effective attraction between two hydrophobic particles.

The hydrophobic interaction does not just explain why oil and water do not mix. Many of the intricate structures that biology forms on the nanoscale are a result of this interaction; it is the dominant driving force that makes proteins fold into well-defined shapes and that causes the soap-like molecules called lipids to come together to form cell membranes. It is an interaction that is not unfamiliar to us at the macroscopic scale—it is, for example, the reason why water droplets do not spread out to wet a greasy surface, but instead form tight balls, minimising the contact between the water and the hydrophobic surface. But at the nanoscale, the hydrophobic interaction can be of overwhelming importance.

Proteins provide an interesting and important case study of nanoscale stickiness. This is obvious to anyone who has tried to clean a pan after boiling some milk in it; the milk proteins form a film that sticks to the metal with a tenaciousness that resists vigorous scrubbing. This poses a serious problem if one wants to implant a medical device into the human body. The lifetime of an artificial hip, or, on a smaller scale, the expanding mesh cylinders that surgeons use to line and open up clogged arteries (stents), may be limited by a sequence of interactions with the body that starts with proteins from bodily fluids landing on their surface and sticking—a process called adsorption. At the smallest scale, the very toxic drugs needed for cancer therapy can sometimes be usefully delivered to the body wrapped up in tiny enclosures called liposomes; their lifetime in the body can be greatly increased if proteins can be prevented from sticking to them. In any medical application of nanotechnology, protein adsorption is going to be a crucial problem that will need to be overcome.

Why are proteins so sticky, and what can we do about it? One problem is that proteins are very heterogeneous. The surface of a protein molecule will have patches with all sorts of different properties. One region will be positively charged, another region negatively charged; all sorts of different chemical groups will be present. So, no matter what the character of a surface with which the protein comes into contact, there will probably be some part of the surface of the protein's molecules that will stick to it. Having stuck to the surface, the protein will quite often unfold—having landed as a fairly compact ball, the molecule will spread out like a paintball fired at a wall. As we will see in the next chapter, one of the main driving forces for the linear protein to take up a globular shape is the need to keep the hydrophobic part of the molecule in the centre of the ball, shielded from the water. Near a surface the hydrophobic part is free to spread out. When neighbouring splattered-out protein molecules come into contact, they stick to each other too, until they form a tenacious film, completely covering the surface.

There must be some way of stopping proteins from sticking indiscriminately, because cells and bodies are full of surfaces. One effective trick seems to be to coat the surface with a hairy layer of water-soluble polymers. For example, drug-containing liposomes can be coated with a layer of the water-soluble polymer polyethylene glycol—such so-called 'stealth liposomes' prove to be much more stable in the body. The mechanism by which these protein-resistant layers work is still a little obscure, but it seems most likely that this is another example of steric stabilisation, working in the same way as gum arabic in stopping the soot molecules in Indian ink from sticking together.

Protein adsorption is not all bad, though; many cooking processes rely on using the adsorbed proteins themselves to stabilise an interface between two liquids or between liquid and air. A frothy foam on the top of a pint of traditionally made beer is testament to the power of proteins to adsorb to the surface of water, unfold, and make a robust film around each bubble.

The mechanical properties of small things

An elephant is different to a scaled-up mouse; a spider scaled up to the size of a horse sounds frightening, but would probably be completely incapable of standing up without its legs buckling. The mechanical properties both of materials and structures can depend on size in quite dramatic ways.

Let us first distinguish between the intrinsic properties of materials and the way you need to build structures, as both of these can depend on size. We can characterise the mechanical characteristics of a material by two properties, stiffness and strength. A material's stiffness tells us how much it will tend to deform when a load is put on it, while its strength gives us a measure of how much load it will sustain before it breaks. They are quite separate properties. One can have stiff materials that are also strong, such as steel, and floppy or elastic materials that are weak, like jelly. But some stiff materials are weak— this is the case for brittle materials, like glass and pottery; and some flexible materials are rather strong—the nylon fibres used in ropes, for example.

How do properties like stiffness and strength depend on size? Firstly, there is the simple observation that a thick beam is stiffer than a thin one, and a thin rope will break under a smaller force than a thick one. But we can define measures of stiffness and strength that factor out this trivial dependence. If we consider, not the force that a wire is subjected to, but the force divided by the cross-sectional area, then we have a quantity called the stress. For metal wires, for example, we would expect that the loads that they could carry before breaking to be proportional to the area of the wire, so we would expect the ultimate tensile stress which the wires would sustain before breaking to be independent of the thickness of the wire; a measure of the properties of the material itself. Likewise, if we consider, not the absolute extension of the wire under a given stress, but the ratio of the extension to the original length (this quantity is called the strain), then we find that the ratio of stress to strain is a constant characteristic of the material, a measure of the stiffness of the material called the Young modulus.

How do these quantities relate to the way in which the atoms are joined together to make the material? In the case of stiffness, the relation is relatively straightforward; in a solid, bonds of various types hold the atoms together. If these bonds are strong and dense, then the material will be stiff. This is the case in a material like diamond, where the carbon atoms are held together by very strong covalent bonds, with the result that the material itself is very stiff. On the other hand, in a material like candle wax, the forces holding the molecules together are the much weaker van der Waals forces, and the result is that the material is relatively easy to bend. The stiffness of a macroscopic lump of material, then, is a direct reflection of the forces that operate on the nanoscale, and as a result of this we expect the Young modulus of a very small piece of matter to be much the same as that of a lump one can hold in one's hand.

The same cannot be said for strength, however. One might think that the strength of a macroscopic lump of material reflects the combined strength of all the bonds that hold the atoms together, but it turns out that this is not the case. Knowing the strength of interatomic bonds, we can calculate how strong a material ought to be on this simple assumption, but what we find is that the strength calculated in this way is orders of magnitude greater than what we actually measure.

So why are materials so much weaker than we think they ought to be? The answer is that materials fail first at their weak spots. This is most obvious for a brittle material like glass. If we want to cut a piece of window pane to shape, we scribe a thin groove using a hard glass-cutter. Then, when we press down on the glass around the groove, it neatly breaks with only the most gentle pressure. We have introduced a weakness at the surface of the glass, and at this weakness only a very small applied force is sufficient to start a crack. Once that crack has started, it very quickly spreads through the material until it breaks.

When we cut glass like this, we are deliberately introducing a weakness. But even if we do not do this, there will always be flaws present. There may be cuts and scratches at the surface of the glass, or microscopic bubbles within it, perhaps; any one of these flaws, when the material is put under stress, may be where the material starts to fail. What determines the strength of a material, then, is not an intrinsic property of the material itself, like how strong the bonds are between the atoms, it is determined by what imperfections, what flaws, the material contains.

This means that strength, unlike stiffness, does depend on size. Something small is less likely to contain imperfections than something big, and if it does have imperfections (cracks, for example) then they are necessarily going to be smaller. For this reason, small things tend to be intrinsically stronger than big things.

This is dramatically illustrated by the case of glass. Glass is a very brittle material, that normally needs to be used with great care in any situation where it needs to carry a load in tension. However, it has been known for a long time now that glass fibres, which have been drawn out to be very thin, can be very strong. The same is true of carbon; graphite, which is used as the 'lead' in pencils, is very soft, but a similar form of carbon, in the form of very fine carbon fibres, is very strong.

A composite material exploits the extra strength you get when parts are small by using thin fibres to reinforce a much weaker 'matrix' material, usually made from some kind of resin or plastic. Glass-fibre-reinforced plastic is familiar to makers of small boats, while the stronger and stiffer carbon fibre composites are used in expensive tennis rackets and golf clubs, as well as in even more expensive military and civilian aircraft. The fibres in these materials have thicknesses of the order of microns. Can we get further advantages by pushing down to the nanometre scale?

Some people certainly think so. They suggest that nanotechnology will yield new materials so revolutionary that they will allow us to perform previously unimaginable engineering feats. Perhaps the most striking of these goals is the idea of building an elevator to space. We would put up a geostationary satellite—a satellite which orbits the Earth above the equator at the height for which its orbital period is precisely one day, so it always remains directly above the same spot on the Earth's surface. Between the satellite and the ground we would string a single cable, up which a cabin—perhaps containing

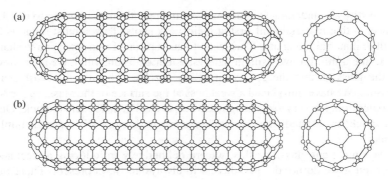

Fig. 4.2 Two types of single-walled nanotubes, in each case by half of a C_{60} molecule. P. J. F. Harris. *Carbon nanotubes and related structures*. Cambridge University Press, Cambridge 1999.

the cargos needed to build and equip interplanetary space vehicles, perhaps carrying the crew of a space station—would climb. This would be very much cheaper than our current methods of blasting stuff up in rockets; in a rocket most of the energy used is needed to carry, not the payload, but the fuel to lift the rocket itself. The only snag is that we need a cable of unimaginable strength and lightness. What could do this? The best candidate is a new, nanoscale form of the familiar element, carbon—the nanotube (see figure 4.2).

Actually nanotubes, despite their novelty, are not so very different from more familiar forms of carbon, and should perhaps be thought of as nothing more than particularly well-bred soot. We all learn at school that the element carbon comes in two quite distinct forms, each of which has very different crystal structures and very different properties. In diamond, the carbon atoms are strongly bonded together in a three-dimensional network; the resulting solid is transparent and very hard. In graphite, on the other hand, the carbon is arranged in hexagons in infinite sheets. Within the sheets, the carbon–carbon bonds are quite strong, but the forces holding the sheets together are quite weak. This results in a solid which is very soft; in fact, because the sheets slip past each other very easily, graphite is a good solid lubricant. As everyone knows, if you rub a piece of graphite over a piece of paper, layers of carbon are scraped off onto the paper leaving a black mark—hence the use of graphite in pencils.

Of the two forms of carbon, it is actually graphite which is most stable under normal conditions of temperature and pressure (hence the fact that diamond is so much more expensive). When we burn a hydrocarbon—such as candle wax, or a plastic-like polyethylene—in conditions where we restrict the amount of oxygen available, the unburnt carbon is left in the form of soot, which consists of incomplete sheets and fragments of sheets of graphite, with a few tens or hundreds of carbon atoms linked together in hexagons, with the carbon atoms on the edges of the fragments, their unsatisfied bonds satisfied by being linked to a hydrogen atom. Similarly, when charcoal was important for other things besides barbecues, the incomplete burning of wood to yield graphitic carbon was an important industrial process. In a modern day variant of this, polymer fibres are charred to yield carbon fibres, which unlike charcoal, are very stiff and strong. This is achieved by making sure that, in the fibre, the chains of carbon atoms are aligned as well as possible along the axis of the fibre. When the fibre is charred, rolled-up sheets of graphite are formed.

In carbon fibre, there is still quite a lot of disorder in the way that the sheets are rolled up to make the fibre. There will be ragged edges of incomplete sheets buried in the piles, and this sort of imperfection will reduce the strength of the fibre from the theoretical value you would calculate for a perfect sheet of graphite. But one can imagine rolling up graphite to get perfect tubes, with no imperfections. Such an object is called a carbon nanotube.

Nanotubes were first discovered in 1991 by Sumio Iijima, an electron microscopist working at the NEC laboratories in Japan, but the story properly begins a few years earlier. In 1985, Harry Kroto, a chemist from the University of Sussex, and Richard Smalley, from Rice University, Texas, had discovered a new form of carbon, in which a graphitic sheet was folded round in three dimensions to make a closed sphere made from exactly sixty carbon atoms. The geometry of this structure, in which twelve hexagons are replaced by pentagons to enable the sheet to fold over and close on itself, is architectural in its symmetry and beauty, and in fact it was named after the architect who proposed using such structures to make domes, Buckminster Fuller. Kroto and Smalley were rewarded for their discovery by a Nobel prize, Buckminster Fuller had a molecule named after him (buckminsterfullerene, more familiarly known as Buckyballs), and the hitherto unexciting world of carbon chemistry suddenly looked full of possibility.

The first problem was to work out how to synthesise this new form of carbon in some kind of quantity. Smalley and Kroto had discovered buckminsterfullerene by vaporising the surface of a piece of graphite with a high-power laser beam. But the quantities of the material they produced were miniscule, sufficient only for analysis by the most sensitive techniques. It was not until 1990 that a paper was published describing how macroscopic quantities of buckminsterfullerene could be made. Wolfgang Krätschmer and Donald Huffman had used a carbon arc to vaporise the graphite in a helium atmosphere, and had found that the resulting soot had a high concentration of buckminsterfullerene and the analogous, ovoid molecule with seventy carbon atoms.

This is where Sumio Iijima entered the story. He had already studied the soot from a carbon-arc evaporation apparatus some ten years previously, finding clusters of carbon in a variety of interesting forms. In fact, in retrospect, one the structures he observed was in fact a nanotube, though at the time its significance was not recognised. It is quite possible that other workers had observed nanotubes before, too, because the evaporation of carbon using an electric arc is a very common procedure in electron microscopy, where thin carbon films formed in this way are often used as supports for other materials.

But after the discovery of buckminsterfullerene, Iijima became very much aware of the significance of his earlier work, and began a systematic examination of the products of carbon-arc evaporation. Soon he was rewarded by the discovery of nanotubes.

Nanotubes have been widely predicted to become amongst the most important components of a number of nanotechnologies. They can be used as ultra-fine probes in scanning probe microscopes, and they may well be useful as components of molecular scale circuits, a subject we will return to in Chapter 9. But it is their strength and stiffness that concerns us here. The hope is that, given that they can be prepared almost entirely free of defects, it may be possible that they achieve almost their full theoretical strength, many times stronger than conventional carbon fibre. We would not expect them to be that much stiffer than carbon fibre, because the stiffness of materials is not much affected by the presence of defects. If we could create nanotubes that were long enough to make continuous fibres, then we could imagine spinning a yarn from them that would be extraordinarily strong—the strongest material available to engineers.

Quantum effects

How would the colour of an object change if we could shrink it from macroscopic to nanoscopic dimensions? This may seem to be a slightly odd question; surely colour is a property of a material rather than an object, and in any case, how can an object which is too small to see have a colour? Nonetheless, the question is more sensible than it seems. What we perceive as colour arises from the way the material interacts with light. Light consists of electromagnetic waves whose wavelength varies between around 400 nm and 700 nm; we perceive different wavelengths as having different colours. White light is a mixture of many different wavelengths, so if we illuminate an object with white light and it only reflects back wavelengths of certain frequencies, then we will see it as being coloured. What determines what kind of light is reflected back depends on two things. The material itself will have a certain response to the light, and this depends on how the light interacts with the electrons within the material. The behaviour of electrons is not described by ordinary mechanics, like the behaviour of macroscopic objects—it is the realm of quantum mechanics. Quantum effects become very important as the dimensions of objects shrink through the nanoscale, so this gives us one way in which size can have the effect of changing an object's colour. But the way the material is structured can also affect the way it interacts with light, through interference effects.

Photonics, interference structures

Suppose that we take a surface and etch a pattern of circular pits in it. If the pits are large, then the surface will look unsurprising—just rough and textured.

But if we make the pits very much smaller—as small as the pits that feature the surface of a compact disk—we see very spectacular effects—rainbows of colour appear as streaks across the surface, seeming to move as we tilt the surface. The colours arise from the phenomenon of interference. Any object that has some structure whose characteristic dimension is similar to the size of the wavelength of light, that is 400 nm–700 nm, will show this kind of effect. In another very simple example, a thick layer of oil on the surface of a puddle of water looks rather dull; it is simply a clear layer of one transparent liquid on top of another transparent liquid. But if the oil layer is thin—thinner than a micron or so, say—then it will show a rainbow of spectacular interference colours. Interference is a consequence of the wave nature of light. A light wave is a pattern of peaks and troughs in an electrical field which travels along at the speed of light. If two light waves originate from two slightly separated places—say from the top surface and bottom surface of a very thin oil film, then the two waves can combine. If the peaks of the first wave fall in the same place as the troughs of the second wave, then the two waves will cancel each other out, but if the peaks of the two waves are in correspondence then the two waves will add up to make a single wave with a greater, combined intensity. As white light is made up from waves with many different wavelengths, each of which corresponds to a different colour, when it is reflected from a thin oil film the waves corresponding to some colours will be reflected, but those corresponding to other colours will not, giving rise to the brilliant iridescent colours that are characteristic of thin films.

One can use interference in thin films to make a perfect mirror. By this I mean a mirror for which all the light is reflected without any of the energy being lost. Mirrors made out of metals can be fairly close to being perfect, but on each reflection a little bit of energy is lost. This matters in something like a laser; here the material that emits the light is put between two mirrors. The light emitted bounces back and forth between the mirrors, causing a little bit more emission each time it goes through and building up a huge intensity of light. Because the light is reflected at each mirror so many times, if only a fraction of energy is lost each time then this all adds up to a considerable total loss. Remember the game one plays as a child (or as a grown-up, for that matter), where one places two mirrors next to each other and looks in, seeing an endless sequence of reflected images stretching away toward infinity . . . only before you reach infinity the images become dimmer and dimmer as well as smaller and smaller, until you can no longer make them out. With a perfect mirror, could you see into infinity?

The nearest we can get to a perfect mirror consists of a stack of alternate thin layers of two materials, both of them perfectly transparent, but with different refractive indices. This ensures that where the two materials meet, light is reflected (as it is, for example, at the surface of water, or at the surface of a piece of glass submerged in water). The thickness of the layers is such that, when the light reflected from each boundary emerges, all of the peaks and all of the troughs exactly reinforce. All of the light coming into the layer is completely reflected, but because both materials are completely transparent none of the light is lost by being absorbed. The only snag is that this only works for a restricted range of wavelengths and viewing angles. But the corollary of this is that, because it is a perfect mirror, it is a material through which light (at least light of that wavelength travelling at that angle) cannot penetrate. It is completely opaque, even though the materials it is made of are completely transparent, simply because of its structure. Such a structure is called a dielectric multilayer, or more grandly, a distributed Bragg reflector.

A spectacular natural example of interference colours is found in the gemstone opal. Fine examples of this stone have beautiful, iridescent colours—yet the material it is made from is simply quartz, crystalline silicon oxide, which in its pure, bulk form is quite clear and colourless. Again, the sole origin of these beautiful colours is the structure of the material on the sub-micron level. To form opal, silicon oxide naturally falls out of a solution in water in the form of perfect spheres, all of which have the same size of about half a micron. The spheres gently settle at the bottom of the solution, and as they settle they pack in a state of near perfect order, like a well-stacked pile of oranges. Then the perfectly stacked spheres are cemented together to yield the solid gemstone. When light falls on the stone, every sphere reflects some of the light, and according to the angle at which you are viewing and the wavelength of the light, the peaks and troughs of the waves reflected from each sphere may either reinforce each other or cancel each other out.

What interests physicists about opal is the possibility that it gives a way of making an even more perfect mirror than the dielectric multilayer which I described above. The problem about the multilayer mirror is that it only works in one direction. But a stack of spheres, neatly piled, looks the same in very many directions—it has a very high degree of symmetry. Could it be that a stack of spheres of the right size could have the property of reflecting all light, without energy loss, no matter what direction it came from? It turns out that opal itself cannot do the trick. But calculations show that there is some form of regular stacking, and some combination of materials that does have this property—in the jargon, it has a photonic band-gap.

Given that we can design such a material, how can we make it? Remember that the characteristic dimensions needed are in the range of lengths defined by the wavelength of light—400–700 nm. This is at the upper end of nanotechnology, and it is an interesting case study in the competing advantages and disadvantages of the two approaches to making things on this scale. In the top-down approach we start with a big lump of material and shape it using the techniques of lithography described in the last chapter. This is the method that has been most successfully used up to now to make artificial structures that have interesting photonic properties. But opal is made by nature using a bottom-up approach, in which sub-micron size components are brought together by a process of self-assembly. It is an attractive approach, because it is much less work than the top-down method. But this application is a demanding one; the stacking has to be close to perfect to get the best results, and the imperfections that self-assembly often introduces are a real problem. Many laboratories around the world are now trying to work out the technical tricks that would make the self-assembly approach work; we will need to wait and see which of the two approaches works best when all the wrinkles are ironed out.

Having talked about a perfect mirror made to exploit interference, how does an ordinary mirror work? This is a good place to start discussing how light and electrons interact, which is what we need to know about to understand the intrinsic colours of materials, rather than the colours that result from their sub-micron structure.

Electrons are charged particles. If they are put in an electric field then they will feel a force. Since a light wave is just an electric field which varies with time, the effect of shining a light on an electron is to shake it back and forth. But if an electron moves backwards and forwards—if it oscillates—then this will create an oscillating electric field—a light wave—which will be re-emitted in the direction from which the original light wave came. A sea of electrons that were completely free to move, and which did not lose any energy when they did move, would constitute a perfect mirror, reradiating any light that reached it back the way it came.

Some good metallic conductors of electricity, like silver or aluminium, are actually surprisingly close to containing, in effect, a sea of free electrons, which is why they make such good mirrors. To understand why this is so, we need to understand some quantum mechanics.

According to quantum mechanics, electrons, just like light, should be thought of as waves. In a conductor, like silver or gold, or in a semiconductor, like silicon, not all of the electrons are bound to circle around the nucleus of a single atom, like those pictures of orbiting electrons that people used to illustrate the atomic age. Instead we should think of the electrons as waves, some of which are able to move all the way through our piece of metal with little hindrance. It is this fact that electrons are delocalised that underlies the ability of metals and semiconductors to conduct electricity at all. A single electron somehow pervades the whole of a block of metal in the same way that the broadcasts of BBC Radio 1 fill the entire nation's airwaves (physics tells you that the cheerful voices of the DJs are present with you, wherever you are, whether your radio is on or not).

Does this mean that there is no interaction between the electrons and the rest of the atom at all? No, it cannot, not least because we know that the electron is negatively charged and the rest of the atom must be positively charged, so there must be a strong attraction between them. What happens is that as an electron goes past an atom, it is scattered; the wave is reradiated circularly outwards. But in a solid we do not just have a single atom, we have a whole crystalline array of them, neatly arranged on a regular lattice. What happens to all the scattered electron waves coming from all of these atoms?

Because they are waves, they interfere with each other. In most circumstances, we have some pairs of waves for which the troughs and peaks add up, but in other pairs they cancel, so when you add up all of the interference from all of the waves scattered from all of the atoms, you end up pretty much with what you started with, a wave travelling through the material largely unaffected by all that scattering. But for some conditions, for electron waves with some particular relationship between the wavelength and the distance between the perfectly arranged atoms, the peaks all cancel with the troughs. We have what to the electrons looks like a perfect mirror, or a perfectly opaque material.

The wavelength of an electron is related to its energy, so what this means is that there are some ranges of energy that an electron cannot have if it is to live inside this block of material. This gap in the range of possible energies is known to solid state physicists as a band-gap.[8]

Now, what happens if a light wave hits our material? As we described above, we would expect it to give any electrons it encounters a bit of a shake. The oscillating electric field exerts a force on the electron, which makes it move—in effect, energy is transferred from the light wave to the electron. But now we have to bear in mind another mysterious feature of quantum mechanics. This tells us that the energy in a light wave comes in little indivisible packets—quanta. Light (like electrons, and indeed everything else, according to quantum mechanics) is at once a wave and a particle. The particles that make up light are called photons, and each photon carries a little packet of energy. The size of the packet of energy is related to the wavelength of the light by inverse proportion, so in short wavelengths the packet is large, and for long wavelengths the packet is small.

This aspect of quantum theory means that when light hits the material, the amount of energy it can give an electron by shaking it up has to be the energy of a single photon, which depends on the wavelength. But what if the electron, if given this extra packet of energy, would end up with a total energy in one of the forbidden band-gaps? This cannot happen, so instead the light must end up passing through the material without affecting the electrons. This, then, is why materials are transparent. Diamond is clear and transparent, because light of all visible wavelengths has photons whose energy, if added on to the energy of an electron in diamond, would take its total energy into the band-gap. Silicon looks grey and silvery, like very highly-polished coal, but if we could see in the infra-red then we would be able to see through it. Silicon has a band-gap which is small enough for visible light to have enough energy so that it can push an electron right over the forbidden zone. Infra-red light, though, with its longer wavelength and lower-energy photons, cannot kick the electrons hard enough to get them across the band-gap. Unable to make any impression, the light passes through unaffected.

Whether light can go through a material or is absorbed by it depends on whether an electron can take up the energy of a photon and still have an

---

[8] One might think that it ought to be called an electronic band-gap, to distinguish it from the closely analogous phenomenon for light that we discussed above, the photonic band-gap. But the concept was developed for electrons long before it was recognised as being important for light, so unqualified, the term band-gap refers to electrons.

allowed energy. The pattern of allowed and disallowed energy states for a given solid is known as its band structure, and it is a major achievement of solid-state physics that this can now be calculated for any given solid, knowing how the atoms of that solid are arranged in its crystal. When you know the band structure, you can deduce all of the optical properties of that solid. We have not yet discussed them, but it also allows you to deduce the electrical and magnetic properties too. But there is another way in which you can affect the allowed states, and thereby alter the ranges of energies that the electrons can have. When the structure and dimensions of the material are controlled on the nanoscale, we can precisely control what energies the electrons can and cannot have, and in this way we can design materials to have whatever optical properties we like.

Although an electron is somehow spread out right through a block of a metal or semiconductor, it is still trapped inside it. The wave corresponding to an electron must be reflected back the way it came whenever it hits the side of the block. In one dimension, this is exactly like the situation when a piece of string is fastened at both ends; a standing wave is formed, and the only waves that are possible are those that have a whole number of half-wavelengths of the wave fitted into the distance between the points at which the string is fastened. This means that the effect of having boundaries is to restrict the number of different types of waves that can exist, just as we saw in our discussion of the Casimir effect. Confinement affects electron waves just as it does any other sort of wave, by restricting the possible values of wavelengths the object can have. If the size of the block is very much bigger than typical wavelengths of the electrons, then this confinement does not matter very much. Electron energies are constrained to take on one of a finite range of values, but there are very many possible values of the energy and they are very closely spaced. But when the block of material enters the nanoscopic domain, things are different. There are many fewer allowed energies available for the electrons, and this is going to have an impact on the optical properties of the material. As we make our block of material smaller, first we see subtle changes in its optical properties—effectively its colour—then, for blocks only a few nanometres in size, we get properties that are quite different from the bulk material (see figure 4.3). By controlling the size of these tiny pieces of material, we can design optical properties to order.

Such tiny specks of material are now known as quantum dots. Many academic laboratories are spending a lot of time studying them, and several university scientists are hoping that their start-up companies are going to make them rich by exploiting the possible applications of these new materials. But it is probable that craftsmen and artists have been exploiting this principle of controlling the size of nanoparticles to control their optical properties for many centuries, even if they did not understand what made it work. In some stained glass, the colour arises from impurities that are dissolved in the molten glass. As the glass is cooled, the impurities gather together in little clusters, that grow very slowly, because of the slow rates of diffusion in the very viscous glass

Fig. 4.3 Vials of cadminum selenide nanocrystals of various sizes between 3 and 5 nm in diameter, irradiated by ultraviolet light, showing the effect of size on the fluorescence colour of the particles. Image courtesy of Felice Frankel, originally in colour as seen on the back cover of this book. Reprinted with permission from B. O. Dabbousi, J. Rodriguez-Viejo, F. V. Mikulec, J. R. Heine, H. Mattoussi, R. Ober, K. F. Jensen, and M. G. Bawendi. *(CdSe)ZnS Core-Shell Quantum Dots: Synthesis and Characterization of a Size Series of Highly Luminescent Nanocrystallites* Journal of Physical Chemistry B **101** (1997) 9463–75. Copyright 1997 American Chemical Society.

melt. Unable to grow to more than a few nanometres in size, these particles form natural quantum dots, whose optical properties contribute to the rich colours of medieval stained glass.

Quantum dots are made today with considerably more deliberation and control. There are two broad approaches. One way starts by making a very thin layer of the material you want to make a dot from. This can be done by sophisticated methods like molecular beam epitaxy or chemical vapour deposition, in which atoms are deposited one by one onto a very flat surface of another semiconductor. Then you heat the surface; just as sometimes when you wipe the windscreen of your car you see first a uniform film which then breaks up into droplets, the thin film of semiconductor will also break up into little drops. The size of the quantum dots can be controlled by how high in temperature you heat the surface and how long you keep it hot. The advantage of making the dots this way is that they are already incorporated in a semiconductor structure. You can incorporate them in structures including dielectric multilayer mirrors of the kind I talked about above, and use them to make new types of light-emitting diodes and solid-state lasers, whose colour is controlled by the size of the quantum dots at their core.

This method does not work if you do not want dots sitting on a surface, but instead want them in solution. Methods of preparing very fine particles dispersed in water are actually quite old; Faraday, in the nineteenth century, devised a method of making a suspension of gold which is deep red in colour, and chemists have been devising methods of making very finely-divided particles ever since. Modern methods make particles of semiconductors like cadmium selenide and gallium arsenide that are very small, have very well-controlled sizes, and whose surfaces are treated to control the forces between them and other surfaces. These have useful properties like fluorescence in colours that depend on the size of the particles, and they are finding uses as labels for the kinds of light microscopy described in Chapter 3.

Before leaving the subject of the behaviour of electrons in small systems, it is worth mentioning that there is another way of thinking about these phenomena. Quantum mechanics teaches us that all particles, including electrons, behave in some ways like waves, and in some ways like particles. We have described the effect of confining electrons in terms of their wave-like behaviour, but there is an equally valid way of thinking about these effects, that predicts exactly the same phenomena, and that considers them as particles.

What distinguishes a quantum particle from a classical particle is that it has to obey a strange set of relationships which derive from Heisenberg's uncertainty principle. There is no problem in principle in simultaneously knowing how fast a classical object is going and where it is. But the same is not true in quantum mechanics; according to Heisenberg's uncertainty principle, the more precisely you know the position of a particle, the more uncertainty you must have about its velocity, and vice versa. In fact, the product of the uncertainty in position and the uncertainty in momentum (that is, the velocity of the object multiplied by its mass) must always be greater than a certain constant. Now, we know that an electron cannot escape from a quantum dot, so the uncertainty in our knowledge of the position of that electron is effectively defined by the size of the quantum dot itself. As we make our quantum dot smaller, then the uncertainty in position is decreased. The uncertainty principle means that, to compensate, the uncertainty in momentum must increase. On average, electrons in smaller quantum dots must be going faster than electrons in bigger quantum dots, and so their energy must be higher. This extra energy, due to the fact that the electrons have been imprisoned in a smaller space, is known as a confinement energy. It is as if the electron, like a tiger in a cage, is content to lie down quietly if the cage is very big. But if it is imprisoned in a smaller cage, then it paces up and down. As the cage gets smaller and smaller, our confined electron goes faster and faster.

'Fantastic voyage' revisited

Physics is different in the nanoworld, and the design principles that serve us so well in the macroscopic world will lead us badly astray when we try to apply them at these smaller scales. The 'Fantastic voyage' submarine may well have worked magnificently at the scale at which it was built, with a full-scale Raquel Welch sitting in the driving seat, but if it was to be shrunk down to fit inside a blood vessel then things would quickly start to go wrong. But this does not mean that it is not possible to design artefacts and devices that do work on these smaller scales. A different feature of the physics that leads to problems for one type of design may be turned to advantage in a design that is properly optimised for this different world.

This is perhaps most obvious in the case of the quantum mechanical effects that occur at small sizes. The changes in electron energy levels that occur when electrons are confined offer us ways to tailor the electrical and optical properties of materials with a remarkable degree of control. Rather than having to take the materials we are given, with the properties nature has assigned them, we can, to some extent at least, engineer in exactly the properties we need.

Brownian motion might seem to be a property of the small systems that is less easy to turn to our advantage. One might think that this constant random shaking that is an inescapable feature of the nanoworld (at least at ambient temperatures) is going to make engineering design very difficult. But maybe we need to think laterally, and try and exploit the opportunities it offers. The speed with which Brownian motion can move things around is a very handy feature of the nanoworld. Because molecules can diffuse so fast over distances of a micron or less, we do not really need to go to any effort to move molecules around. We should be able to throw away many of the pipes and pumps we need to transport fuel around, for example; the fuel molecules will find their own way from the tank to the motor quite fast enough.

The stickiness of the nanoworld that arises from the strength of surface forces on small scales also looks like it is going to be a big problem. But again, this feature can provide many opportunities for the ingenious nano-designer. Brownian motion and strong surface forces combine to lead to a powerful approach to fabricating devices which only works on the nanoscale. If molecules are synthesised with a certain pattern of sticky and non-sticky patches, the agitation provided by Brownian motion can lead to the molecules sticking together in well-defined ways to make rather complex nanoscale structures. The key to understanding this mode of assembly—known as self-assembly— is that all of the information necessary to specify the structure is encoded in the structure of the molecules themselves. We will explore this thoroughly in the next chapter.

Where should we look for models for a good design that is well adapted to the special features of the nanoscale? The answer is all around us and within us. Living organisms are made up of cells, and cells are full of sophisticated machines that have evolved over billions of years to operate in the peculiar (to us) environment of the nanoscale. To a human engineer, the materials and workings of the cell can look ramshackle and poorly thought out. The materials seem inadequate; the main structural elements are floppy, carbon-based polymers and even more transient assemblies of detergent-like molecules— little more than water-filled soap bubbles, in fact. The designs seem positively baroque, with fantastically involuted surfaces and bizarrely-shaped aggregates of molecules. Surely, our engineer thinks, if we sat down with a blank piece of paper and redesigned the thing from scratch we could do better.

A view that is more common than it should be is that nature has done the best she can in the face of the difficult constraints imposed by evolution. I would argue the opposite; at the sub-cellular level, nature's designs are in fact highly optimised. Only our experience of designing things in a realm in which physics is very different prevents us from recognising this.

5
Making soft machines

Successful scientists are just as aware of the power of an image as any advertiser or marketer. The image that has, more than any other, defined the promise of nanotechnology was made in 1989 by Don Eigler; appropriately it depicted the logo of a company—IBM—picked out in individual xenon atoms (see figure 5.1). To make the image, individual atoms were moved across a surface into place using a scanning tunnelling microscope, an instrument that was invented in IBM. Enthusiastic disciples of extreme nanotechnology have seized on this image as proof that their goal—the manufacture of nanoscale machines an atom at a time—is now in sight. I think this is wrong. These experiments are a magnificent technical achievement, and they have allowed scientists to study some fascinating new physics. But they are not a step on the path to the development of true nanoscale machines, because they encourage us to think about making these machines in ways that do not take into account how different physics is at these very small scales. It is all too easy to visualise a cantilever that picks up an atom and sets it in its place as being just like a robot arm that picks up some component in a car factory and bolts it onto the engine. This kind of analogy between the macroworld and the nanoworld is both seductive and dangerous. Thinking about nanoscale design in this way leads to the risk that our designs will not work. This is because those factors

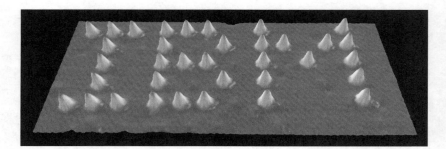

Fig. 5.1 The IBM logo picked out in individual atoms. Xenon atoms on a nickel (110) surface. See D. M. Eigler and E. K. Schweizer. *Nature* **344** (1990) 524–6. Courtesy: IBM Research, Almaden Research Centre. Unauthorized use not permitted.

that are unimportant at the macroscale, like the continuous random jiggling of Brownian motion, become crucial and limiting at the nanoscale.

But not only is there a risk that we will use design principles that are wrong or inappropriate to the nanoworld because of our lack of intuition and familiarity with the different physics that operates there, but also that same lack of intuition may lead us to neglect to use the very different, but very powerful, design principles that are appropriate for the nanoworld. The most important of these design principles is the subject of this chapter: self-assembly. In self-assembly, the information needed to make a three-dimensional structure is contained within the molecules themselves, rather than having to be imposed from outside. We do not have to place the molecules one by one according to our blueprint. The blueprint, if it makes sense to talk about one at all, is contained in the molecules, and all we have to do is to gather the molecules together for them to make the structure themselves.

Before dealing with the right way to make objects at the nanoscale, let us consider in more detail why ways that derive from our macroscale intuition may not work at the nano-level. For example, in some popular visions of nanotechnology, the central enabling concept is the molecular assembler. We imagine a machine with a tiny robotic arm, small enough and precise enough that it can pick up individual atoms and molecules and place them at will. We further imagine that the assembler has a tiny on-board computer that controls the movement of the robotic arm and the placement of the atoms. Thus, we will be able to program the assembler to produce whatever assembly of atoms we like. Of course, this includes other assemblers; the first thing we can do once we have built our first assembler is to program it to build some more assemblers, each of which will build still more assemblers. When we have got enough assemblers we will be able to take any source of simple atoms and build anything we like from them. This could be a perfect copy—atom for atom—of the most exquisite artefact, or (as assemblers will be available in limitless quantities) we will be able to use them to make everyday items from their component atoms; perhaps an apple or a sandwich. Thus, in the view of the nanotechnology disciples, we will enter a millennial period of unparalleled plenty.

The IBM experiment proves that, in principle, we can pick up an atom and place it with some precision. The cantilever of the scanning tunnelling microscope can indeed be thought of as a robotic arm that can position atoms precisely. Why then is the goal of a molecular assembler, which would represent the scaling down in size of such a robotic arm, not a good idea? One reason is because in the light of the different physics at the nanoscale, many barriers need to be overcome for the machine to work in practice. At the nanoscale, remember, Brownian motion becomes more important. The robot arm will be subjected to such a barrage of molecular collisions that molecular precision becomes difficult. In Chapter 3, we saw how, even at the relatively large scale of an AFM cantilever, this random bombardment leads to a random motion of the tip on a scale not much smaller than the atomic scale; as we attempt to shrink the robot arm yet further we will inevitably end up with a less stiff structure that will deform much more in response to the Brownian forces.

These random forces are not the only forces that we have to worry about. We know that, on the nanoscale, surface forces become very important. Our robot arm will develop an alarming tendency to stick to things, and having stuck it will be hard to get it free. Our intuition about machines and robots for constructing things is developed in the macroscopic world, and is an unreliable guide to the nanoworld. At this point, some would argue that these objections underestimate our ingenuity in designing engineering solutions. If stiffness is a problem, can we not find some stiffer material to work with—diamond, for example? If stickiness is a problem, can we not coat our components with some non-sticky material, like the fluorocarbon coating of a non-stick saucepan? We could work this way, and we would probably make some progress. But would it not be better if we could find a way of designing that made a virtue out of the lack of stiffness and the stickiness, and the continual random motions of the nanoworld? There are design principles that do just that, and we know they exist because they are the principles used by nature. After all, an apple does pretty much assemble itself anyway, without any need for molecular assemblers.

The design principles used in cell biology actually rely on those features of physics in the nanoworld that we, with our intuitions honed in the macroworld, think of as problems to be overcome. In self-assembly, we rely on the importance of surface forces at the nanoscale to stick things together. We depend on the continuous random jiggling of Brownian motion to assemble the components and bring them together in the correct orientation for them to best fit together. We need our components to be floppy rather than stiff, so they can adjust their size and shape to suit the structure that they are making. The benefits of using self-assembly are that, if we design and synthesise the right molecules, then all we have to do is to throw them together in a container to make our complex three-dimensional structure. The process is massively parallel rather than serial. With our robot arms, a single robot can only make one object at a time. With self-assembly, we control how many structures are made by how many molecules we put in the pot. Instead of making objects one at a time, we can make billions of billions of them in one go. Self-assembly is self-correcting. If, by chance, one of the structures is put together incorrectly, then Brownian motion will shake it apart and it will have another chance to put itself together properly. It is also self-healing. If some external shock breaks the structure up, then it can reform itself without any help from outside; this is possible because of the way in which the information needed to make the structure is carried in the molecules themselves. Many of these properties remind us of the properties of living things. This is because self-assembly is a fundamental part of how living things operate. We should not overstate this; there is much more to life than self-assembly. But equally, because self-assembly is the best way of building things at the nanoscale, it is not surprising that these are the principles that biology has exploited and developed to a high degree of sophistication. Plenty of synthetic systems show self-assembly too, so there is nothing in principle to stop us from borrowing the methods used at the cell level in biology and copying them using artificial molecules. Indeed, I will argue that this will be the design method of choice for nanotechnology.

Self-assembly

Most people, when they are trying to solve a jigsaw puzzle, like to look at the picture on the box. This picture provides the blueprint for what the solved puzzle should look like; we solve the puzzle by deducing what small part of the large picture each piece represents, and we check whether our deduction is correct by seeing whether the piece fits snugly where we think it belongs. The more advanced puzzler may do without the picture, but even if he does not have a complete blueprint to help him solve the puzzle then he still relies on outside information; that the blue pieces are most likely to belong at the top of the picture, and parts of faces are likely to be oriented with the eyes above the nose.

Is it ever possible to assemble a jigsaw puzzle without a picture to provide a blueprint, or without using any intelligence or information that is not intrinsic to the pieces? Of course it is, if the jigsaw is carefully designed. To see this, imagine a jigsaw puzzle the shapes of whose pieces are designed in such a way that any piece will only fit in its proper place. This jigsaw can be assembled in a way which is completely mechanical and completely reliable. We start with one piece and select another piece at random. We try all possible ways of fitting the two pieces together; if the two pieces do not fit then we discard our trial piece and select another piece at random, carrying on in this way until we find a good fit. When we have our two pieces fitted together, we select a third piece at random, and again try all combinations of the two-piece kernel of the finished puzzle with the random piece. Once again, if there is no good fit then we discard this piece and try some more. This completely random procedure allows us reliably, but laboriously, to assemble the puzzle.

Where was the blueprint for the finished puzzle? For this kind of jigsaw, where it is impossible to fit a piece into the wrong place, there is no need to have a picture on the box to follow. Instead, the information that is needed to assemble the puzzle is fully contained in the shapes of the pieces. The procedure for assembling the jigsaw is completely automatic, and no more information needs to be used to put it together perfectly. The process of self-assembly can be compared to this kind of jigsaw. We imagine molecules that have complementary shapes, that fit together to make some kind of three-dimensional shape. In one spectacular example from biology, certain viruses are made from protein and RNA molecules; the shape of the protein molecules is such that all we have to do is put the molecules in a test-tube and the protein pieces will spontaneously fit together to form the rather spectacular shape of the viruses.

The jigsaw analogy for self-assembly is popular and useful. It is very important to realise, though, that it is incomplete, and it can be misleading if we use it too blindly without considering the different physics that operates at the nanoscale. What it does succeed in capturing is the idea that the information that contains the design of the assembled object is not contained in some external blueprint, but instead is encoded in the components of the object themselves. What the jigsaw analogy is less good at conveying is the fact that self-assembly relies on some mechanisms that specifically depend on the fact that physics is different at the nanoscale. The pieces of the jigsaw need to be thought of as being in continuous random motion, due to the effect of Brownian motion, while we should think of the connections between the pieces, less as being complementary shapes, and more as being the result of patches of stickiness.

What characterises a self-assembling system is a certain balance in strength between the effect of Brownian motion, that tends to shake things apart, and the size of the surface forces, that tends to stick them together. We can push the jigsaw analogy a little further to see the importance of this balance. Imagine a child's toy consisting of a rabbit made out of pieces which fit together in a three-dimensional jigsaw. To help the process of putting the rabbit together, we coat the faces of the blocks that will be in contact in the final shape with a weak glue. Now we put all of the pieces of the rabbit in a bag and give it a shake. What happens? It depends on how sticky the faces are relative to how violently you shake the bag.

If the shaking is relatively gentle, and the faces are very sticky, then as soon as any pair of faces come into contact, whether they fit each other well or not, they will irreversibly stick. The structure that results will be very irregular and its precise shape will depend on the contingencies of chance—how exactly the pieces fell into the bag, and the precise way in which you carried out the shaking.

If the shaking is much more violent, and the faces are not very sticky, then then no permanent structure will be formed at all; no sooner than a contact between two sticky faces has been made than it is shaken apart again. But for some happy medium of stickiness and violence of shaking, the pieces will stick enough to form the structure; but if the two faces that stick do not quite fit then or if they meet at something less than the perfect angle, then they can unstick themselves enough to find a better fit. After a little shaking, we will be able to pull out of the bag a perfectly formed toy rabbit.

For self-assembly to be a dominating factor, we need the right balance between the effect of the surface forces, that tend to make components stick together, and the Brownian motion, that tends to pull them apart. In practise, these tendencies generally come into balance on length scales between the molecular and the micronscale, and that is why self-assembly is so effective as a design principle for the nanoworld.

The most striking examples of self-assembly in our everyday world come from soap and the soap-like molecules that make up shampoos, hair gels, cosmetics, and the other potions and lotions that seem to be essential for modern life. What these materials have in common is a class of molecules known as amphiphiles—one part of one of these molecules has an affinity for water, and is referred to as hydrophilic. The other part of the molecule is hydrophobic—the hydrophobic parts of the molecules tend to cluster together to avoid being in contact with water. To resolve these two opposing tendencies—to maximise the amount of contact between the hydrophilic parts of the molecules with water, while minimising the amount of contact between water and the hydrophobic parts of the molecules—a mixture of an amphiphile and water will take up quite complex and ordered structures. These structures can have striking mechanical properties, despite the fact that they are still mostly water.

Self-assembly is taken to its most sophisticated form in biology. The basic structural unit of life are proteins; these are linear molecules that for the proper operation need to fold into a well-defined three-dimensional shape. The resulting object has both a complex shape and a pattern of sticky and non-sticky patches on its surface that allow it either to interact with other molecules that fit into the shapes on the surface, or to come together with other protein molecules to build up the complex machinery of the cell.

Before moving on to discuss these examples in detail, we need to consider more deeply the role of Brownian motion, and the disordering tendency that it implies, in how self-assembly works. The tendency of things to become more disordered, which Brownian motion acts as the local agent of, represents a fundamental physical law—the second law of thermodynamics. Yet here we are arguing that self-assembly leads to the spontaneous production of ordered structures. This apparent paradox needs to be resolved by a careful consideration of the nature and meaning of disorder, or in more technical language, of entropy.

Order from disorder

C. P. Snow famously said that it should be a mark of any cultured person, whether or not they were a scientist, that they should know the second law of thermodynamics at least as well as the works of Shakespeare. But of all the fundamental laws of physics I suspect that the second law is the least understood, even though statements of it, often expressed in non-technical terms as 'the disorder of the universe always increases', are widely known. These misunderstandings can cause even well-educated people to hold major misconceptions, for example, the idea that the theory of evolution is inconsistent with the second law.

The apparent paradox behind self-assembly—the idea that the tendency to universal disorder can lead to the spontaneous order—goes to the heart of the problem. The second law of thermodynamics needs to be considered in the context of statistical mechanics—that branch of physics which deals with the properties of large groups of interacting objects (for example, atoms and molecules) in a statistical way, without attempting to predict the detailed motion of the individual molecules. It is in these statistical terms that we need to try and understand what is meant by disorder. If we look at a glass of water, for example, we now know that it is made up of many molecules, all of which are in a state of continuous, random motion—Brownian motion. Nonetheless, we can describe properties of the glass of water—the volume of water, for example—which remain unchanged even while all the molecules are moving around in their random and chaotic ways. In terms of statistical mechanics, there are many different possible arrangements of the water molecules, all of which produce the same macroscopic state. At the microscopic level, the arrangement of water molecules is constantly changing; the microscopic state of our glass of water (which we could specify in principle by listing the positions and velocities at any instant of each of the molecules in the glass) is continuously changing.

How do we know what macroscopic state we should expect to find the glass of water in? By simple probability; if one macroscopic state is consistent with the largest possible number of microscopic states, then that is the state we will find it in. We can say that the glass of water is in a state of maximum disorder, where we can precisely define the degree of disorder as being the number of different microscopic arrangements of the system consistent with a given set of macroscopic properties.

We can illustrate this with an everyday example. Supposing I were a tidier person than I actually am, and I kept my collection of CDs carefully filed, with all of the classical music in one section, in alphabetical order by composer, and all of the rock music in another section, filed by artist. This is a state of low disorder, because there is only one possible microscopic state, consisting of the carefully-filed arrangement of CDs, consistent with the macroscopic state of all of the slots in the CD rack being filled. If I have a party, and all of the rock CDs are pulled out, to be replaced in random slots in the rack the morning after, then my collection has moved to a state of higher disorder; if we do not respect the alphabetical order of artists then there are many ways to arrange the rock CDs in a way in which the racks are still full (precisely, we can say that if there are N distinct rock CDs then there are $N!$ ways of arranging them on the rack). Now, if my baby daughter systematically empties all of the racks and I carelessly replace them completely at random, mixing up the rock and the classical CDs, then the collection is in a still greater degree of disorder, as measured by the still larger number of distinct arrangements of CDs that are possible. In this example, the natural tendency of systems to move toward a state of maximum disorder is easily understood. Physicists measure disorder by a quantity called entropy. This is closely related to the measure of disorder given above in terms of the number of possible microscopic arrangements.[9]

There is an intimate relationship between the idea of entropy and the idea of information. If the arrangement of something is completely random, then it

[9] In fact, if the number of arrangements is defined as Ω, then the entropy is given by $S = k_B \ln \Omega$. Using the logarithm of Ω rather than Ω itself ensures that the total entropy of two independent systems is simply given by adding up the entropy of each system; the constant k_B—Boltzmann's constant—effectively provides a conversion between the units used for energy and the units used for temperature.

cannot store any information. If I need to decrease the degree of entropy of a system, then I need to somehow put in some information. The process of sorting, by which I might separate out the classical CDs from the rest, needs an input of information, and the result is a system in a lower state of entropy than it was before.

The second law of thermodynamics states that, for an isolated system, the entropy is always a maximum. So how, if the disorder or the entropy is always to be maximised, can an ordered state spontaneously form? The answer arises from the fact that, although for a system as a whole the entropy is maximised, this need not necessarily be the case for a smaller part of the system. In fact, we can pay for a decrease in entropy in one part of the system by an increase in entropy, at least as big, in another part of the system. If we are to say about any system that its entropy must always increase, then we must make sure that the system is isolated from its surroundings, so that this kind of trading of entropy from one system to another cannot take place.

But most systems that we encounter are not isolated from their surroundings—in our glass of water, molecules could leave the surface and join the atmosphere as water vapour, and heat can flow from the water to the glass. In this case, we need to imagine an impermeable, insulating box around the water and the surroundings; it is in this combined system of the water, the glass, and the surroundings that the total entropy must be a maximum. What this means is that within the box we can trade entropy; it can be possible for order to appear in one part of the system (say the glass) at the cost of increasing the amount of disorder elsewhere.

Suppose we decrease the temperature of our glass of water and its surroundings below zero degrees centigrade. The water in the glass will freeze. The ice is certainly in a state of lower entropy, lower disorder, than the liquid water. In the liquid the molecules could be anywhere in the glass, while in the ice they must occupy definite positions on a crystalline lattice—therefore there are many more possible arrangements of the water molecules in the water than the ice. To achieve this higher degree of order locally, we must effectively export entropy into the surroundings. The way this happens is that when the water freezes it releases heat—the latent heat of freezing—which has to flow into the surroundings before the ice crystal can grow. It is this export of heat which effectively increases the disorder of the surroundings. So, within the framework of the second law of thermodynamics, we can achieve spontaneous local ordering by exporting both heat energy and entropy to the surroundings.

To keep track of these flows of energy and entropy, physicists and chemists use a composite quantity called the free energy. This is defined as $F = E\text{-}TS$, where E is the energy, T is the temperature, and S is the entropy. The importance of this quantity is that it allows us to restate the second law in a way that is more convenient to use; it turns out that, for a system that can exchange energy with its surroundings, the condition that the entropy of the system and surroundings together is a maximum is equivalent to the condition that the free energy of the system alone is minimised. This idea of free energy lets us make more precise the idea that to get self-assembly one needs a balance between the stickiness of the interactions that hold the self-assembling components together and the Brownian motion that tries to shake them apart. The energy part of the equation is minimised if we maximise the number of sticky contacts between the components, but maximising the entropy favours less-ordered arrangements with fewer sticky contacts.

The balance between energy and entropy that leads to the lowest free energy depends on the temperature; at low temperatures we expect more-ordered structures, while at higher temperatures disordered structures may be favoured, even if they have a higher energy. Self-assembled structures, then, are dynamic structures which result from a precise balance between order and disorder. We made an analogy between self-assembly and a jigsaw puzzle, in which the molecules are the pieces which fit together to make the picture. We can now see why this analogy is incomplete, because it concentrates on the order and leaves out the disorder. To make the analogy accurate, we would have to imagine the pieces of the jigsaw to be flexible, and constantly in motion. Every now and again a piece becomes disconnected from its neighbour, only to flip back into place again a little later. Occasionally, a piece will jump right out of the jigsaw, only to be replaced a little later by another piece. Self-assembled structures are soft, because their components are continually in motion, rather than being firmly locked into place. Yet these soft structures are also highly robust and even self-healing.

Soap

Some of the best examples of self-assembly to be found outside biology can be found in the bathroom and kitchen. Soap (and the soap-like molecules known as detergents or amphiphiles), when mixed with water in various proportions, can form structures with an extraordinary complexity of structure. The physical properties of these solutions range from the relatively ordinary—fluid solutions of soap or detergent which only betray any special character by a slight haziness—through the familiar but intriguing gloopy and jelly-like qualities of thick shampoos and shower gels, to the downright weird phenomenon of a clear gel which looks like water, but is actually rigid enough to ring when it is tapped.

To understand what is at the bottom of these unusual physical properties we need to study the arrangement of the soap molecules within these solutions. Such studies (carried out by neutron or X-ray diffraction, as described in Chapter 3) reveal that some of these solutions are a bizarre mixture of the fluid and the solid, in which soap molecules take their place in complicated arrangements of soap and water molecules. In these structures the overall architecture can be patterned with an exquisite degree of order and symmetry, yet each molecule still remains fluid, ceaselessly popping in and out of the structure as Brownian motion keeps both water and soap molecules in constant motion.

It is an ambivalence about its molecular character that makes a soap molecule so effective at forming these soft, self-assembled structures. One end of the molecule finds a watery environment quite congenial—it is hydrophilic (it is usual to call this the head of the molecule). The rest of the molecule (the tail) has the character of oil, which would very much prefer to be in the company of other oily molecules rather than being forced into contact with water molecules. If only a very few of these soap molecules are present in a solution with water, then they will remain separated from each other, but as we add more soap there comes a point when rather than remaining isolated little clusters of individual soap molecules—perhaps 50 or 100 at a time—will come together to form what are known as micelles. If we were able to look closely at these clusters, we would find that the water-loving parts of the soap molecules were concentrated on the outside of the cluster, in contact with the water, while the oily parts form a central core, shielded as much as possible from the water by the hydrophilic groups around the outside.

What shape do these clusters or micelles form? It depends on the shape of each individual soap molecule (see figure 5.2). A molecule with a relatively large area of head and a tail of relatively small volume can be thought of as a little cone, and an assembly of such cones will pack together to make a sphere. This is the usual situation with true soap and with many detergents. On the other hand, a molecule with a particularly bulky tail compared to its head size can be thought of as a cylinder; these will pack together to make, not a sphere, but a sheet two molecules thick, with the head groups lining the face of the sheet next to the surrounding water. In practise, the easiest way to get such a bulky tail is to have, not one, but two tail groups attached to each water-loving head; this is the molecular design of molecules called lipids, which are centrally important to biology as the main building blocks of membranes. There is a rarer, intermediate type of molecule which can be thought of as a truncated cone; these will come together in the form of giant, flexible cylinders, which are rather graphically known as 'worm-like micelles'. Although such solutions are less common, their properties are very striking. If one tries to stir the solution fast then the

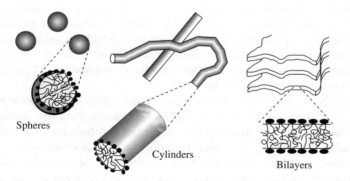

Fig. 5.2 Some of the shapes of aggregates that can be formed by soap-like molecules.

worms get tangled up with each other and dramatically impede the flow, but, if the solution is gently poured, then the micelles have time to get out of each others' way, and the flow is much more liquid-like.

The picture we have so far is rather close to the jigsaw picture of self-assembly—we imagine our molecules as solid pieces with sticky patches which stick together in certain well-defined ways. Molecules with different shapes can be thought of as leading to structures with more or less natural curvature; molecules whose shape implies a lot of natural curvature will pack together to form spheres, while molecules whose shape implies less or no natural curvature will form sheets. This is a useful way to begin thinking about the problem, but unsurprisingly it has its limitations; it is too static a picture, and neglects the all-important roles of Brownian motion and entropy. Their effects show up in the striking changes in shape that happen when we have a micellar soap solution and we add more soap molecules.

Supposing that we have a soap solution in which the micelles are spherical. As we add more soap molecules, rather than going freely into solution, they make more micelles. The density of these micelles increases, until a significant fraction of the total volume of the solution is occupied by them. At this point, the way in which the spherical micelles can best pack together becomes important. The most efficient way to pack together an array of spheres is an ordered packing. If you think of stacking oranges, then you get the most oranges in a given volume by packing them in triangles and hexagons. The next layer sits happily in the pockets created between three oranges on the plane below, forming another plane of hexagonally-packed fruit. This kind of packing is known as 'close packing', and it takes only quite simple geometry to show that, when perfect spheres are packed in this way so that they touch, the spheres take up 74% of the total space. Randomly-packed spheres, on the other hand, fill space less efficiently, with a maximum packing fraction of only 63%. The consequence of this is that, as we increase the density of the micelles by adding more and more soap to our solution, at some point we get an abrupt change from a situation in which the micelles are random, to one in which they are packed in an ordered array.

It is worth thinking harder about the origin of this transition from disorder to order, because it illustrates, yet again, some of the apparent paradoxes that underlie the whole phenomenon of self-assembly. What drives a solution with a high volume fraction of micelles from a state in which the micelles are completely disordered into a state in which the micelles are ordered is the fact that the ordered state has a higher entropy than the disordered state. How can this be, if entropy is a measure of disorder?

We measure the state of disorder in a precise way by asking how many different microscopic ways of arranging the system are there that produce the same macroscopic appearance of the sample (i.e. a certain fraction of the volume being occupied by the micelles, either in a state of disorder or in an ordered array). Measured in this way, we find that some counter-intuitive trade-offs can happen. Some entropy is associated with the degree of order of the packing of the micelles, while another part of the entropy is associated with how much room each individual micelle has to move around. In our densely-packed mass of spherical micelles, by arranging the spheres in an ordered array, each individual sphere gets more room to move around, and this extra opportunity for disorder for an individual sphere outweighs the loss of disorder implied by the regular array of spheres.

The limits of the jigsaw picture of self-assembly become most obvious if we take a soap solution in this state of an ordered array of spheres and add even more soap to it. Geometry tells us that there is an ultimate limit to how much space can be filled by packed spheres of 74%. But the spheres are not hard objects; they are soft and mutable assemblies of soap molecules that can change their shape in response to a change in their environment. This is exactly what they do if we increase the concentration of soap; we find an abrupt change from a state in which the soap molecules pack in spheres to one in which they pack in long, cylindrical rods. Remember that we have assumed that the shape of the individual molecules is such that they would rather pack as spheres. What drives them to pack in this way, which from the point of view of the individual soap molecules is not ideal, is once again the consequences of packing the micelles. Closely-packed spheres can fill no more than 74% of space, but closely-packed cylinders can fill up to 91% of space. In another

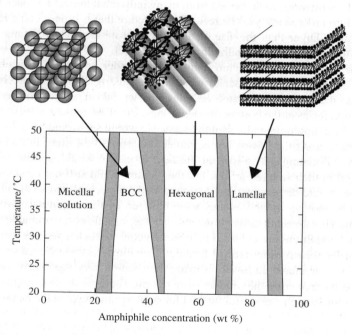

Fig. 5.3 Phase diagram for solutions of a simple, uncharged soap-like molecule in water. The material is a short triblock copolymer of ethylene oxide (EO) and propylene oxide (PO), with the structure $(EO)_{37}(PO)_{58}(EO)_{37}$. BCC is the body centred cubic phase. Data from P. Alexandris, D. Zhou and A. Khan. *Langmuir* **12** (1996) 2690. Reproduced from R. A. L. Jones, *Soft Condensed Matter*, Oxford University Press, Oxford (2002).

trade-off, individual molecules will pack in a non-ideal way so that the resulting micelles have a shape which gives them more room to explore.

We can go further and imagine adding even more soap to our solution, which by now will be rather a pasty gloop. How can we pack yet more soap molecules in, while still giving the aggregates of molecules room to move about in? The answer is, yet another shape change. The molecules become packed, not in cylinders, but in sheets, with each sheet composed of two layers of soap molecules, with the head groups facing the water outside and the oily tails in the centre of the sheet. The advantage of this arrangement is that, while the maximum fraction of space that can be filled by packed cylinders is 91%, sheets can be packed together to fill up to 100% of space.

These most-concentrated soap solutions, consisting of stacks of layers of assembled soap molecules, are in fact what bars of soap are made of. The characteristic slippery, 'soapy' feel of a bar of soap when one fumbles for it in the bath comes from the ease with which these layers slip over each other. The case of the bar of soap illustrates much of what is subtle and complex about self-assembly.

The process of self-assembly can be hierarchical. At the lowest level, individual molecules pack themselves in structures which reflect both the shape of the individual molecules and the locations of their sticky patches. At the next level, the structures, each formed from many individual molecules, pack themselves into ordered arrays. The result is a structure that is ordered on a length-scale much larger than the size of an individual molecule or the range over which the forces between individual molecules act.

The self-assembled structures, despite their order, are soft. This is obvious at the macroscopic length scale; what we are talking about here are bars of soap, at the most-concentrated end, and, as we add more water, the slimy gloop that forms at the bottom of soap dishes. But if we looked at these materials at the nanoscopic scale, then the scene we would see would be one of constant motion. At the lowest levels, molecules would be flitting in and out of micelles. Zooming out to look at the next level, we would see the micelles arranged in their ordered arrays, but the picture would still not be static. The micelles would be bending and flexing as molecules came and went, and they would be moving around as they were buffeted by the continuous Brownian motion. This constant activity means that the structures are mutable; if the external conditions are changed then the structure will change in response.

Finally, the structures are self-healing; if we impose some external stress that breaks the structures up, say by stirring the solution very vigorously, or if we remove the stress, then the structures will reform. These are the features that make self-assembly such a powerful method for creating structures at the nanoscale.

From shoe soles to opals

The basic units that come together in self-assembly do not have to be ordinary small molecules, as they are in soaps. What we need for self-assembly to work is for the surface forces, that stick the units together, and the Brownian motion, that shakes them apart, to be roughly in balance. This will tend to happen for units whose size varies from the molecular (roughly a nanometre) up to a few microns. So our self-assembling units can be small molecules, but they can also be the very large molecules known as polymers or macromolecules.

As we will see, macromolecules are the building blocks of biology, which relies on their self-assembling properties. Simpler, synthetic macromolecules are what plastics are made of, as well as being the major components of all kinds of glues and gums. While most commodity plastics, like polyethylene and polypropylene, are too simple to show very much in the way of interesting self-assembly behaviour, slightly more complicated molecules known as block copolymers can show a rich range of self-assembled structures. These structures have fascinated academic researchers, not least because the theory needed to understand them is rather easier than for soaps, but they have also found a few everyday uses.

As long as they are not too big, the basic units of self-assembly do not have to be single molecules. Any solid particle which is small enough for Brownian motion to be important for is known as a colloid, and these particles too can self-assemble. Emulsion paints and water-soluble varnishes are made of suspensions in water of tiny particles of plastic, each made up of very many individual polymer molecules (see figure 5.4).

The gemstone opal is made up of tiny spherical particles of quartz. Whenever we have spherical particles, all of which are approximately the same size as each other, we can expect the same transition from disorder to order that we saw with spherical micelles. At low concentrations of particles, they will be moving around completely randomly, buffeted by collisions with

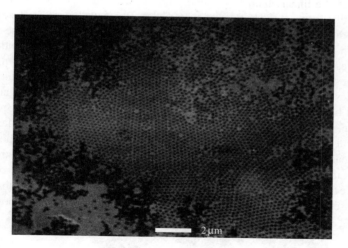

Fig. 5.4 A water-borne latex suspension of the kind that might be found in a water-based varnish, imaged with an environmental scanning electron microscope. Note the ways in which the particles form regions of ordered packing. Reproduced with permission from C. He and A. M. Donald. *Langmuir* **12** (1996) 6250. Copyright 1996 American Chemical Society.

molecules of the liquid they are suspended in. But as the concentrations get higher, the fact that spheres pack more efficiently in space when they are in an ordered array than when they are disordered means that we get a transition, driven by entropy, to a state in which the particles are packed in a crystalline array. It is this self-assembled ordering of the particles that gives the opal its beautiful, iridescent appearance.

In opal, the particles are similar in size to the wavelength of light, so light waves scattered from neighbouring particles produce interference patterns, in the same way as light being reflected from the front and back of an oil film on water produces spectacular colours which change according to the angle from which you look at them.

The simplest self-assembling macromolecule is really just a soap molecule scaled up in size. A diblock copolymer consists of two chemically-different linear macromolecules joined together at the ends. If one of the blocks is soluble in water (let us call it the 'A' block) and one (the 'B' block) is not, then in solution this material will form micelles, just like a soap molecule. What makes these molecules interesting is that even without the addition of water they will form complex, yet ordered, structures.

What is the driving force that leads to these structures? It depends on the nature of the interaction between the A and B blocks (see figure 5.5). For most combinations of chemical units, there will be a slightly unfavourable interaction between an individual A segment and a B segment; an A segment would on balance prefer to be in the company of other A segments than B segments. But because we are considering a macromolecule, it is not just one A segment that we are dealing with; we have a large number of A segments joined together in a linear chain.

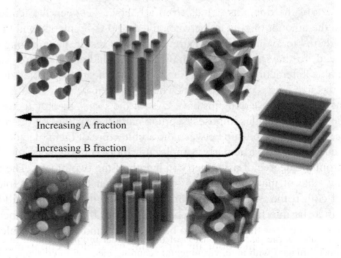

Fig. 5.5 Block copolymer phases. Graphic courtesy of James T. Hoffman, The Mathematical Sciences Research Institute. Copyright James T. Hoffman.

The number of segments can easily be hundreds or thousands. If one A segment has a slight preference for the company of other A segments, then a hundred or a thousand A segments will have an overwhelming urge to shun the B segments, because the interactions are additive. So if we have a mixture of A macromolecules and B macromolecules then they will have a strong driving force to separate completely, like oil and water. But in a diblock copolymer, no matter how much the A blocks and B blocks want to get apart from each other, they cannot; we have chemically joined the chains. A divorce of the A blocks and the B blocks is not possible, and the best they can do is to arrange to have separate bedrooms.

A neat block copolymer usually consists of nanoscopic domains of A and B, separated by an interface. How big are the domains? On the one hand, the two types of blocks would like to minimise their contact, and clearly the bigger the domains the less contact they have. On the other hand, because the two blocks of the copolymer are joined, if the domains are too big then each block would have to stretch to fill it. Why should a macromolecule not want to stretch into a more or less straight line? Once again, the answer comes from considering the different physics on the nanoscopic scale. Think of a macromolecule as a long piece of string, but one in which every segment of the string is being bombarded by its neighbours and is in continual random motion. Stepping back, we would see the string in a continual state of flexing and folding. If we were able to grasp a single macromolecule by one end and pull it straight, then we would feel a force trying to pull the end back. The origin of the force is the random collision of the molecules surrounding the chain causing it to bend and buckle; if we let go of the end then it would retract and the chain would fold up into a random tangle which had the maximum amount of disorder. We would be observing directly the tendency to maximum entropy implied by the second law of thermodynamics; this tendency is so real that it produces a physical, measurable force.

The size of the domains comes from a balance between the tendency to minimise the amount of contact between the A and B segments, and the tendency of chains to avoid being too stretched. As a result, having long chains would lead to having larger domains, while having a more strongly unfavourable interaction between A and B would lead to smaller domains. Usually the size comes out to be somewhere between 10 and 100 nm.

What shape do these domains have? As for the soaps, the key factor is that the shape of the molecules imposes a certain natural curvature on the resulting structure. But the relationship between the architecture of the molecule and the natural curvature is particularly simple for block copolymers—it derives simply from the degree of asymmetry in length of the two blocks. The natural state—the state of maximum entropy—of a polymer chain is as a folded-up tangle, or coil. If the two blocks are of the same length, then they will be satisfied with a flat interface, but if one block is bigger than the other then there will be a tendency for the interface to be curved around the smaller block. So we would expect sheet-like domains for symmetric diblock copolymers.

As we increase the degree of asymmetry we should have a transition, first into cylinder-shaped domains, and then to spherical domains. These are indeed the most common structures, and they have been studied in diblock copolymers for nearly thirty years. More recently, some more exotic morphologies have been discovered. One of the most intriguing is the so-called 'gyroid' phase. Here, rather than having discrete domains, we have two, interlocking, continuous regions of A and B.

The ingenuity of polymer chemists is such that we do not have to be restricted to block copolymers made up of two different blocks. We can have linear polymers with more than two blocks. With three blocks linked together, for example, we can have either an ABA structure, or we can have three chemically-different blocks to give us an ABC structure.

An interesting feature of ABA triblocks is that the two A blocks on a single chain can end up in different domains. If we have spherical A domains, as will

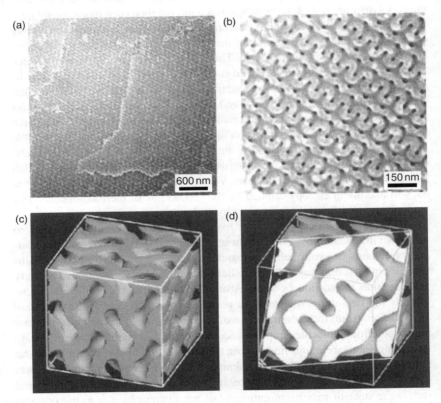

Fig. 5.6 A complex, self-assembled structure in a block copolymer of polystrene and polyisoprene diluted with pure polystrene. (a) and (b) are scanning electron micrographs of the structure at different magnifications, while (c) and (d) are computer simulations. Reproduced with permission from T. Hashimoto, K. Tsutsumi, and Y. Funaki. Nanoprocessing based on bicontinuous microdomains of block copolymers: Nanochannels coated with metals. *Langmuir* **13** (1997) 6869–72. Copyright 1997 American Chemical Society.

happen when the A blocks are much shorter than the B blocks, and at room temperature the A domains are hard and glassy, then these domains will act as links which join the triblock chains together to make an infinite network. A network like this is like a rubber, except that in a rubber the chains are joined together by permanent chemical bonds. In a triblock the links between the chains are reversible and can be removed by heating the material up to a temperature at which the A domains soften. This material is effectively a thermally-reversible rubber, a very useful substance which can be used, for example, to make shoe soles.

Neither are we restricted to simple chains. We can join more than one chain together at a point to make a star, and we can graft a series of side-chains onto a longer backbone molecule. We can mix together different copolymers with different architectures, and we can mix the copolymers with their corresponding simple polymers—it is not difficult to see that endless complexity is possible.

The three-dimensional structures that form by self-assembly from these complex mixtures of architectures are themselves complex and intricate (see figure 5.6). What is important is that, complex though the three-dimensional structures are, the information that they embody is contained solely in the molecules from which they are made. From the size of the chains, their architecture, their asymmetry, and the interactions between their segments, we can predict with considerable accuracy the structures they will form. The molecule itself carries the code for the structure it forms. This idea of a molecular code for three-dimensional structure is immensely important, because we now know that this is the key to the way life itself works.

Self-assembly and life

Anyone with access to a computer and a connection to the internet can go to one of a number of websites and download the complete human genome—the fundamental blueprint of humanity. But what does it mean to call the genome a blueprint—how does this digital string of symbols get converted into flesh and blood? One thing is clear—the genome is not a blueprint in any kind of literal or conventional sense. There are no engineering drawings of what a hand should look like, or how the tubes get connected to the heart, or (to take an example at the cellular level) what the shape should be of the endoplasmic reticulum, a fantastically convoluted surface involved in protein and lipid synthesis which is contained in every cell.

The information in the genome is of a very restricted kind; it simply specifies the order in which the units called amino acids are to be joined up to make a protein. The genome is divided up into individual genes; each of these genes is nothing more than the recipe for making one particular protein. So, the immediate physical realisation of the information in the genetic code is not the cell or the organism, but the collection of protein molecules that the genome codes for—nowadays this collection is fashionably called the proteome.

What would happen if we collected together all of the proteins that the genome of a human encodes, put them in a bag, and gave them a shake? Would a human being emerge? Of course not. The collection of proteins defined by the proteome are not the blocks from which the organism is built, they are the tools with which it is made. Human beings are not self-assembled. In contrast, some viruses can be formed by pure self-assembly; if one puts the components of the virus—some proteins, some DNA or RNA—into a test-tube then they will spontaneously form themselves into virus particles which are absolutely identical in terms of virulence and other properties to viruses that emerge from their host.

But the fact that humans, and other organisms, are not viruses and are not self-assembled does not mean that self-assembly plays no part in the life of higher organisms. It does, but it must do this in a framework in which it is continuously taking in entropy from the environment. A living creature is very definitely not at equilibrium, so, looking at the organism as a whole, it certainly would be quite wrong to think of it as the result of a self-assembly process.

But if we look at a small part of the organism, this may well be heading toward some local state of equilibrium. Even on the cell scale, the processes of life depend on a continuous flow in of energy and entropy, but if we look at a single macromolecule or a small group of macromolecules, then we can often define a local region which is heading toward equilibrium, and in which structures are being formed by pure self-assembly mechanisms.

The best way to keep track of all this is to focus on the idea of information. Recall our definition of self-assembly as being the formation of a three-dimensional structure which is completely specified by the molecules that go into it—where the code for the structure contains no more information than the components it is made from. How do we know when self-assembly stops? The answer is, when we have to put any extra information into the system in order to get the structure we want. The typical processes that would involve an input of information include sorting and keeping things separate in compartments.

But the fact that in some processes we do put in extra information does not mean that self-assembly becomes totally irrelevant. Suppose that we have a mixture of molecules that all together would not self-assemble into anything useful, but which, if we separated them into two different mixtures, would produce two different self-assembled structures that when combined produced a useful machine. Overall, we cannot describe the machine as being self-assembled, because the structure is not completely coded in its component molecules. However, we cannot make the machine without self-assembly. We can divide up the process of making it into three stages.

Stage one involves sorting the molecules into the two groups, and keeping them in compartments isolated from each other. This step requires the input of external information. In the second stage, the two sets of molecules, safely isolated from each others' interference, separately self-assemble into the two structures. In the third stage, which requires another input of external information, we remove the two structures from their compartments and put them together to make the machine. This example might sound slightly far-fetched, but I believe that it is conceptually very close to how most processes at the sub-cellular level operate. I also believe that this is going to be the route by which we make working nanoscale machines and devices.

Protein folding

All living things—including ourselves—depend on the operation of complex molecular machines in order to operate. These are the machines that convert our food into energy, that in plants make light into food, that build the components we need in order to be able to grow, that make our muscles move, and that allow us to reproduce. The molecules from which these machines are built are proteins, and the construction of the machines relies on the fact that proteins are able to self-assemble into very specific three-dimensional shapes.

A protein is a molecule made from a long sequence of smaller subunits called amino acids. Twenty chemically-different amino acids are used by nature to build proteins; each protein molecule consists of a linear chain with an absolutely precise sequence of amino acids. If you think of the twenty amino acids as letters, then every different protein can be thought of as a (rather long) word defined by the sequence of the letters. What is remarkable is that the linear sequence of letters in the word is sufficient to specify precisely the three-dimensional structure of the protein. The process by which the linear protein molecule folds into its working structure—protein folding—is a process of self-assembly.

Before we consider in detail how this folding process works, let us backtrack to the point at which proteins themselves are made. The information needed to specify a protein is a linear code, and in living things this code is archived in DNA. DNA is another linear molecule with a precise sequence of subunits, called bases. In the case of DNA, there are only four chemically-distinct bases rather than the twenty different amino acids, so each amino acid must be coded for by a combination of three bases. (Of course, there are 64 different combinations of three bases chosen from four, so we have more combinations than we need to specify the twenty amino acids. Some of the excess combinations are used effectively as punctuation marks, while in other cases we have more than one combination of bases coding for the same amino acid.)

Every living cell contains in its DNA the specifications for creating all of the different proteins the cell needs in its lifetime—this set of specifications is known as the genome, and the specification for a single type of protein is known as a gene. This point is so fundamental that it is worth stressing. It does not really make sense, strictly speaking, to speak of a gene for blue eyes or long legs, much less a gene for intelligence or an introverted personality. The gene specifies the linear sequence of amino acids in the protein; this linear sequence in turn specifies the three-dimensional shape the protein takes up when it folds; this folded shape dictates how the molecular machine that the protein is a component of operates. It is only the complex of interactions between all of the molecular machines that make up the cell; the other biochemicals that they help synthesise, and indeed the control that some molecular machines themselves impose on the degree to which any given gene is read and converted into proteins, result in the higher-level characteristics of the organism that we observe.

It is that step when the linear protein molecule self-assembles into its well-defined shape that is central to the whole process of converting a simple linear code into a complex organism. How do we know that a protein molecule forms a well-defined structure, and how do we know that that structure is uniquely specified by its linear sequence of amino acids? The structure of proteins can be determined by X-ray diffraction, which we discussed in Chapter 3. Any protein that can be crystallised (and this still excludes many important proteins, particularly those associated with membranes) can be analysed in this way, which produces a list of coordinates of each individual atom in the macromolecule with a resolution well under a nanometre; by now structures are available for thousands of different proteins.

If we take a dilute solution of a protein, and either heat it up or expose it to certain chemicals (denaturants), then we find that its structure changes. The overall size of the molecule increases, certain structural elements weaken or disappear, and any chemical activity the protein had goes away. The well-defined, compact, three-dimensional shape of the protein is replaced by a more open, much less well-defined structure, and as a result of this the molecule can no longer do its job.

That we can rather easily destroy what is a soft and delicate molecular machine is not surprising; what is remarkable is that if we lower the temperature again, or remove the added denaturing agent, then the protein molecule folds back up into exactly the shape that it started as, and it does its job just as well as it did at the beginning of the experiment. It is the fact that the unfolding of a protein is reversible, and that an unfolded protein will refold in an isolated solution, away from any of the machinery of the cell, that makes us believe that no extra information beyond the one-dimensional sequence of the protein is required to specify its working, folded shape. Thus, it is an experimental fact that the sequence of amino acids specifies a single, folded, three-dimensional shape, known as the native state.

It is not at all obvious, though, that it should be that way. Think of a simple diblock copolymer, and let us suppose that the 'A' block is soluble in water at high temperature, but insoluble in water at low temperature. The 'B' block is soluble in water at all temperatures. If we have a diblock copolymer like this in a dilute solution in water, it will effectively have a folding transition with temperature. At high temperatures, the whole molecule will be soluble in water, and will form a random, open structure. As we lower the temperature the 'A' block will want to minimise its contacts with water; it will collapse into a compact globule, with the B segments on the outside of the globule in contact with water.

A protein is more complicated than a diblock copolymer. Instead of having two different types of segment, A and B, we have twenty types of segments, A to T. Whereas, in a block copolymer, the segments being arranged in blocks, so we have many A segments all joined together, and then in another section of the molecule all the B segments in a block, in a protein different types of segment are juxtaposed apparently at random. Nonetheless, we can roughly divide up the twenty amino acids into a group of ten which are hydrophobic (which we call H segments), and another group of ten which are more or less hydrophilic (we will call these P segments, for 'polar').

So the simplest way of thinking about a protein is as a sequence of H segments and P segments; the stable state of the protein is the one in which the number of contacts between H segments and the surrounding water molecules are minimised. This can be achieved if the protein collapses into a ball in such a way that all of the H segments are found in the middle of the ball, and the P segments are on the outside.

So, proteins and soap seem very similar so far. A compact, globular shape is dictated by the need to keep the hydrophobic parts of the molecule away from water. The resulting three-dimensional structure—a single-molecule micelle, probably roughly spherical in shape—is specified by the sequence and interactions of the units. But for soaps and simple copolymers, the three-dimensional structure is not unique. If we look at a large number of copolymer micelles, they will have the same overall plan, with a spherical blob of A in the middle surrounded by B. But if we look at the location of one specific segment—say the tenth A unit in from the end—then we will find a different location in each micelle. This is quite different from a folded protein, in which the position of every segment relative to every other segment is precisely specified.

There is something special about proteins which gives them the power to convert the information in their linear sequence of segments into a shape which is absolutely precisely specified. In our synthetic molecules, we can achieve this conversion after a fashion; we can design a sequence that will give us some control over shape, but only in a vague and imprecise way—we can design a molecule that will produce a vaguely blob-shaped object rather than a more open ball of string.

What is it about proteins that allows the conversion of the linear sequence code into three-dimensional shape to be so much more precise? In a protein, the twenty different amino acids do all have different characters, and our binary division of them into hydrophilic and hydrophobic is too simplistic. Instead of keeping track of three different kinds of interactions, between H and water, P and water, and H and P, we have all of the combinations between A to T and the water as well. In principle, what we could do to find out what shape a protein will take up is to list all of the possible ways that we can arrange the protein chain. We can then count up all of the different interactions between the different segments—we would find this many A–A interactions, this many A–B interactions, and so on, and, knowing the energy of each kind of interaction, we could calculate the total energy of that arrangement of the protein. We can then look through our list of arrangements and find the one with the lowest total energy; we expect that this will be the native state. In reality, as the size of the protein chain that we are studying approaches realistic lengths, the number of arrangements we have to list becomes prohibitively large, but as computers get more powerful it is possible to do this kind of calculation for longer and longer chains.

This kind of computer modelling has led to an important insight that lets us get to the heart of the difference between synthetic systems and living ones. If we generate random sequences of segments, then we find for the overwhelming majority of sequences that the lowest-energy states are compact and globular. But there are many arrangements, which in detail are quite different from each other, which have very similar energies. In the competition to find the lowest-energy arrangement there is not a clear winner. How is it, then, that a real protein can unerringly find the one correct arrangement that corresponds to its native state? The answer must be that the sequence of segments in a protein is not truly random, despite the fact that protein sequences have no obvious regularities. Out of all of the possible sequences of segments, only a vanishingly tiny proportion have the property that there is a single well-defined native state which has the lowest energy. Most sequences produce a situation like a soap or a copolymer, where a large number of interchangeable, globular states can all coexist.

The difference between how nature works to make proteins with a single well-defined three-dimensional architecture and how our current synthetic chemistry operates, achieving at best the ability to make a crude blobby shape, is that nature has at its disposal the power of evolution. We will turn to the way in which evolution can select those molecules which are best able to self-assemble to make useful structures shortly. But first, we need to consider the other major functional components of life, the nucleic acids.

Nucleic acids

The computer age has changed both the way in which we live and the way in which we do business; it has also changed the way we think by introducing some seductive and powerful metaphors. The distinction between hardware and software is one of these metaphors. We think of hardware as physical, in principle capable of directly impacting on the world around, but useless without the direction of software. By itself, this software is entirely ephemeral and non-corporeal, but when its instructions are executed it can animate the previously inert hardware to carry out the tasks we want.

In biology, it has become common to think of the software of an organism as being carried in the genetic code. This code has a physical expression in the molecule DNA. The primary functions of DNA are in information storage and reproduction, but we should not forget that it is a physical object with some unusual and striking properties. The way it stores information relies on these properties, especially for an example of self-assembly of a particularly precise type.

The clever part of the hardware of an organism consists mainly of proteins; as we have discussed, these fold into very specific shapes from which the machines that carry out the functions of the organism are made. In addition to this 'clever', functional hardware, organisms also have a certain number of purely structural elements, like bone, skin, and tendon in animals, which are made partly out of proteins but also partly from other types of molecules.

Biology has one further class of materials that stretches a little the distinction between software and hardware. These are the molecules of RNA; chemically very closely related to DNA, they form the link between the software world of DNA and the hardware world of proteins. Their position is intriguingly ambiguous; they read and carry the information from the DNA to the machines that physically realise the genetic code by creating new protein molecules according to the specification in the genome, but they can also act as machines themselves. RNA molecules, in a sense, are both software and hardware, and this dual role has suggested to many people that, before life evolved to its current stage, a previous, more primitive form of life existed in which RNA served both to store genetic information and to make the machines that would execute that information.

The discovery of the way in which genetic information is stored in DNA is intimately tied to the discovery of the structure of DNA molecules. Each DNA molecule is a polymer. Like proteins, the chain is made from a sequence of different possible subunits, but where for proteins there are twenty possible subunits, DNA is made up by linking subunits of four different types. Each subunit consists of a common backbone, with one of four possible chemical groups hanging off the side. These groups are known as bases, and they are conventionally referred to as A, T, G, and C (see figure 5.7). This set of four bases have a property known as base-pairing, which makes possible their particularly special type of self-assembly. Each of the four bases is a flat molecule, rather like a tile. The arrangement of chemical groups at the edge of the tiles means that, in one particular orientation, three strong bonds will join together the base C with the base G. Similarly, the bases A and T can be held together by two strong bonds. We really can think of these as pairs of jigsaw pieces that, if we bring them together, edge to edge, in one particular orientation, will snugly fit together.

A molecule of DNA can be thought of as a backbone that the bases hang off. If two molecules of DNA have the property that, wherever one base appears on one chain, its complementary base appears on the other chain, then the jigsaw pieces on the two chains will snugly join together. The geometry of this works out such that when this happens the two chains come together in the famous double-helix structure (see figure 5.8). This is a very special and very specific type of self-assembly. It requires absolute precision in terms of the sequence required to create the match. The self-assembled structure—the double helix—is highly robust. These are exactly the properties needed for the DNA

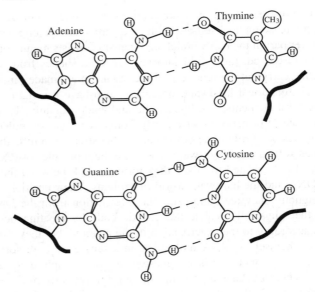

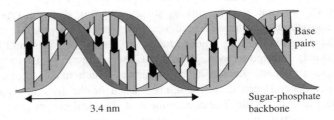

Fig. 5.7 The four bases A (Adenine), T (Thymine), G (Guanine), and C (Cytosine) of DNA, shown as complementary pairs. The thick lines represent the attachment to the sugar-phosphate backbone of the strand, while the dashed lines represented the hydrogen bonds which hold the complementary pairs together.

Fig. 5.8 The double-helix structure of DNA.

molecule to archive information in a robust and stable way. The RNA molecule shares the base-pairing mechanism with DNA. (The base T in DNA is replaced by a closely-related base, U, in RNA, but U binds to A in just the same way as T does.) However, rather subtle differences in the chemistry of the backbone between RNA and DNA mean that RNA does not form a stable double helix. This makes RNA less suitable as a pure information archive than DNA, but underlies its dual role as both software and hardware.

RNA is formed on pieces of uncoiled DNA. When the DNA helix opens up, the base pairs break apart. It is the matching pieces from the box of RNA subunits which are temporarily paired with the exposed DNA bases. Then the

RNA subunits are joined together to make a new molecule of RNA; when enough of the DNA bases to make the code for a single molecule of RNA have been read, then the RNA molecule breaks away, and ultimately forms the template to make a new protein.

Until recently, it was thought that the role of RNA was essentially passive. In this view, RNA was simply a vehicle for carrying the information from the place where that information was archived—the DNA—to the machinery that converts the information into the hardware of a protein molecule. But it is now becoming clear that RNA molecules can by themselves act as molecular machines—as hardware. It is the base-pairing mechanism that allows them to do this.

Even though two RNA molecules cannot come together to make a stable double helix that runs the length of the whole molecule, if, on a single RNA molecule there are two regions in different parts of the molecule where there are matching runs of complementary bases, then these will come together to make a small local section of double helix. Supposing, for example, we had the run of bases UAUUGACCA, and a little further along the chain the run UGGUCAAUA occurred. There would be a tendency for these two complementary pieces of chain to come together in a short section of double helix, effectively gluing a loop into place. We can imagine, then, in an RNA molecule subjected to the random buffetings and flexings of Brownian motion, that the molecule would be trying out all kinds of different shapes in a way that effectively searched for this kind of complementary runs of bases. When two such runs come together, they will stick, and if the run is long enough then they will stay stuck. In this way, the molecule will find the outline of a three-dimensional structure, which will then, perhaps, be stabilised by further, weaker interactions between different parts of the molecule. In this way, the sequence of bases along an RNA molecule can also define a three-dimensional structure.

In principle, RNA can act as a machine, as a piece of hardware. The number of known roles for RNA-based machines in organisms is growing all the time. This recognition that RNA can play a dual role, as both hardware and software, has also led to a growing suspicion that an earlier form of life was completely based on RNA, with proteins and DNA becoming involved only later. We will return to this point briefly when we talk about how molecules can evolve.

Living soft machines

Some proteins perform their function as single molecules—their well-defined three-dimensional shape allows them to act as catalysts, facilitating some chemical change that would otherwise take place too slowly to be useful in the living cell. But many other protein molecules have a further level of self-assembly.

First of all, by the self-assembly process we talked about in the last section, they fold into their native state. But these building blocks can then combine with other proteins and other structural elements to form larger structures—machines—often with complicated moving parts—that perform many of the essential functions of life. Some of these larger machines are themselves the product of self-assembly, again in the strict sense that the components themselves contain all of the information needed to put together the larger structure. But for some of these complex constructions, simple self-assembly is not enough. By learning how nature puts together components to make machines like this, we can learn to go beyond self-assembly, to extend the principles of self-assembly to allow us to add extra information to the building process, while still respecting the special features of design at the nanoscale—the stickiness of surfaces, the continuous jiggling of Brownian motion, and the lack of stiffness.

As we will see, going beyond self-assembly does not mean going back to macroscopic design principles; we still need to respect the different physics at the nanoscale. What are the components of these larger-scale soft machines? Proteins are the key functional elements, and in a few, important machines—particularly those ones that are concerned with making more proteins—RNA is another important component. The ribosome, for example, which is the machine that adds on amino acids to a growing protein chain according to the sequence recorded in the genome, consists of more than fifty individual protein molecules as well as several RNA molecules.

The three-dimensional shape of each protein molecule and each RNA molecule is defined by its sequence, and then the shapes of each of these folded molecules and the distribution of sticky patches on their surfaces further dictate how they all come together to make the working machine.

In addition to these two classes of molecules—proteins and RNAs—with their highly-sophisticated design, a class of smaller molecules called lipids are also of vital importance in making working soft machines. Lipids are the biological equivalent of soaps and detergents, and like these they are simple, rather small molecules, not much bigger than the molecules of soap or detergents. Like soap or detergent, one part of the molecule has an affinity to water—it is hydrophilic—while another part of the molecule resembles oil, and is hydrophobic. The difference between a lipid and a soap is that, while the hydrophobic part of a soap molecule is a single, shortish (perhaps a dozen carbon atoms long) hydrocarbon chain, a lipid molecule generally has two such hydrocarbon chains. As we discussed in the section about soap, this difference in molecular geometry changes the way in which the molecules aggregate in water. Just as for soap, the lipid molecules come together or aggregate in such a way that the hydrophobic parts of the molecule are hidden away from any contact with the water, shielded by the hydrophilic ends. But the roughly cylindrical shape of lipid molecules means that the preferred self-assembled structure is a sheet-like bilayer, with the molecules stacked two deep, the hydrophobic tails in the centre of the sheets and the hydrophilic ends on the outside.

This propensity to make lipid bilayers is tremendously valuable, because it allows living things to make compartments. Over a long distance, if a sheet is flexible enough, it can curve round on itself and its edges can join up to form a barrier that completely separates inside from outside. In a solution of pure lipids, this kind of closed shell is called a liposome.

In fact, the presence of a defensible frontier between the uncontrollable outside world and the ordered interior is so fundamental to living things that it is almost part of the definition of life. All living things are made from cells, and what forms the basis of the outside wall of any cell is a lipid bilayer. In the cells of all but the simplest class of life—bacteria—there are many subdivisions, many organelles, all defined by their own walls, also formed from lipid bilayers.

To make an organism, or even an organelle within a cell, it is not sufficient simply to have a barrier cutting off the organism from the outside world. There must be traffic across the frontier, and that traffic has to be very carefully controlled. Nutrients and foodstuffs need to be brought into a cell from its environment, and waste products must be expelled. The cell also needs to be able to process information about its environment; a bacteria may need to be able to detect a food source so it can swim toward it, while in a multi-cellular organism, such as a human, the different cells communicate with each other by issuing a complex pattern of chemical signals that are detected at the outside of a cell, but which trigger a reaction on the inside of the cell. The agents of all this transport and communication across these lipid bilayers are soft machines made out of proteins.

These selective pores and pumps which move molecules in and out of the different compartments defined by the lipid membranes are based on a particular class of proteins known as membrane proteins. Membrane proteins, like ordinary proteins, have a precisely-defined three-dimensional structure which is defined by the sequence of amino acids along the chain. But membrane proteins cannot fold into their working structures by themselves—they need to be associated with a lipid bilayer.

Recall that our crude picture of a folded protein was a blob in which the hydrophobic segments were buried in the centre of the molecule, shielded from contact with the surrounding water by a screen of hydrophilic segments which make up the outside layer of the blob. A similarly crude picture of a simple membrane protein would show a cylinder in which the hydrophilic segments were concentrated on the two circular ends of the cylinder. The curved surface of the cylinder would be made up of hydrophobic segments. If the length of the cylinder matches the thickness of a lipid bilayer, then, if the cylinder crosses the bilayer, then the hydrophobic segments on the cylinder's curved surface will be in contact with the similarly hydrophobic oily tails of the lipid.

In this way, the self-assembled structure depends on combining information encoded in at least two types of molecules—the shape of the lipid molecules and especially the length of the tails define how thick the bilayer is, while the sequence of amino acids in the protein makes sure it folds to a shape that precisely fits into the bilayer. Without the lipids, the protein sequence does not make sense, and the protein by itself will not be able to fold into a useful, well-defined structure.

We can imagine a further development of our simple membrane protein. Imagine that, instead of forming a simple, solid cylinder, the cylinder has a hole drilled though the middle, and the walls of the hole are lined by hydrophilic segments. Now we have a molecule that, in combination with a lipid bilayer, will act as a pore. A little water-filled tube, stabilised and kept open by the favourable interaction between the hydrophilic segments lining the tube and the water molecules, crosses the lipid bilayer, allowing molecules in solution to cross the barrier.

How can we design a molecule that, in combination with a lipid bilayer, would fold to such a shape? One common motif in nature is to have a molecule a little like a synthetic multi-block copolymer. One has a series of blocks of hydrophobic segments, each block exactly long enough that if it folds up into a helix then the resulting hydrophobic rod exactly spans the thickness of the bilayer. These hydrophobic rods are joined together by hydrophilic segments that will form the caps of the cylinders. The precise sequence of amino acids will control the details of the shape of the folded membrane protein. Most importantly, the sequence will control the size of the pore. In this way, we can have pores that will only transport molecules of a certain size.

Even more complicated structures can be made from assemblies of several proteins in association with a lipid bilayer. One of the most beautiful and important biological machines is ATP-synthase, which is fundamental to the way in which all living things, from the simplest bacteria to multi-cellular organisms such as ourselves—manage their energy. At one level, ATP-synthase is a membrane-bound protein complex which controls the transport of one particular type of chemical—hydrogen ions—across a membrane. But bound to the membrane complex is another complex of proteins; together these form a remarkable machine. When hydrogen ions pass though the pore in the bilayer made by the membrane-bound part of the machine, they cause one part of the machine to rotate with respect to another part. The mechanical energy of this rotation is used to synthesise the energy-storing molecule ATP, which is the power source used for virtually all of the processes of life.

The story of how this machine can first convert chemical energy to mechanical energy, and then mechanical energy back to a different form of chemical energy is a fascinating one that involves some more of the surprising features of the nanoworld. We shall return to this in the next chapter, but first we need to return to the question of how such complicated nano-machines can be made.

How complicated a structure can you make by pure self-assembly? On the one hand, it is clear that self-assembly is sufficient to fold a protein into its correct three-dimensional structure (at least if the protein is in a dilute solution). Some quite complex machines—the ribosome, for example—which themselves are composed of many different molecules, are also self-assembled. If the components are separated, and then put together again in the test-tube, then the structure will reform itself and work as it did before. Even whole viruses can self-assemble; if the necessary proteins and nucleic acids are put together then fully-infectious and pathogenic viral particles will emerge.

Viruses, though, cannot be considered as living organisms—they are information parasites which need to commandeer the metabolism of a host cell to reproduce. Self-assembly by itself is not sufficient to make even the simplest living cell. If we simply took the chemicals that make up such a cell and put them in a test-tube then they would not spontaneously come together to make a functioning organism. In these systems that cannot be put together by self-assembly, what new construction principles are needed?

What is most important to realise is that, even if self-assembly is not sufficient by itself to construct a particularly complicated nanoscale machine, the new principles that need to be used must still respect the special features of the nanoworld. The constant jiggling caused by Brownian motion and the stickiness caused by the strength of surface forces still dominate. This means that we should look for principles that do not replace self-assembly, but instead extend it. How nature does this, and how we can emulate this synthetically, we will talk about next.

Beyond simple self-assembly

An inexperienced or slovenly cook will make a dish by putting all of the ingredients together in one pot. This works fine for some dishes—a stew or a pot au feu, for example. But a more skilled and careful cook knows that one can extend the variety of dishes that can be made, and greatly improve their tastiness by making the dish in stages. If you put flour, butter, lemons, sugar, and eggs into a bowl, mix them up, and cook them then you will end up with a pretty nondescript mess. But if you separately make pastry from the flour and the fat, meringue from the egg white and sugar, and a lemon filling from the lemons and egg yolk (with a bit of butter and sugar) then you can make a delicious lemon meringue pie.

What is crucial here is that the sorting process, which distinguishes a recipe of even modest complexity from the one-pot method, is a way of putting in extra information above and beyond the simple list of ingredients. Different sequences of sorting and mixing can produce quite a number of quite different dishes from the same set of ingredients.

The purest form of self-assembly corresponds to one-pot cooking. All of the information required to make the product is encoded in the molecules themselves, so necessarily it cannot matter in which order the ingredients are put together. In physicists' language, the outcome of the process is an equilibrium situation, which does not depend on the history of the system. We can go beyond equilibrium by introducing a dependence on the history of the system, by introducing extra information over and above what is implicit in the ingredients themselves at various stages of the recipe.

What this analogy does is to suggest ways in which we can retain many of the virtues of self-assembly, while extending its potential to making more complex and varied structures. One of the simplest approaches to extending self-assembly is by sorting—instead of putting all of the ingredients together and mixing, we separate the ingredients into one or more different groups. Each group of ingredients can then self-assemble without interference from the ingredients in the other group. The extra information is put into the system by the sequence of mixing.

Putting in the extra information requires an extra input of energy. We can separate the ingredients either just by ensuring they start out a long way from each other, or we can physically enclose one set of ingredients in a container, releasing it later. A nice example of this principle is found in nature in what are known as chaperones—special protein machines that are used to help other proteins fold to their proper three-dimensional shape. Although an isolated protein will refold to its proper shape, if it is in the presence of a lot of other unfolded proteins, rather than each protein individually refolding without interfering with each other, they may instead stick together in a useless, insoluble mass. This is exactly what happens when you cook an egg—the proteins in the egg white unfold and then stick together in a random and unstructured way, forming the gel-like texture of the cooked egg. When proteins misfold in this way in a living organism, the consequences can be very serious. It is believed that a number of fatal diseases, including Alzheimer's disease, Parkinson's disease, BSE (bovine spongiform encephalopathy), and its human version Creutzfeld–Jacob disease, all originate from protein molecules which, instead of folding by themselves to their proper three-dimensional structure, form insoluble clumps in which many protein molecules are stuck together.

Chaperones are one of the mechanisms by which nature ensures that proteins properly fold. One of the best-studied chaperones—GroEL—is itself a protein molecule that folds up into the shape of a cup with a hinged lid. The way it works is simple; an unfolded protein molecule finds its way into the cup, and the lid is closed on it. The protein, safely kept away from other unfolded proteins with which it might form improper associations, folds to its proper three-dimensional shape and is then released. Note, once again, that the act of sorting one protein molecule out from its neighbours and keeping it isolated while it folds needs an input of energy, which here is used in the cycle of shutting and opening the lid of the chaperone molecule.

Another way of making a structure that could not be made by pure self-assembly is by using self-assembling precursors. This is a particularly useful trick for making robust, insoluble structures, both in nature and in synthetic chemistry. One of the drawbacks, of course, of self-assembled structures is that by definition they are not very permanent. The molecules are brought together in solution and held together by relatively weak bonds. One way round this is to make a structure from self-assembled components, and then subsequently do some chemical processing that makes the structure much more difficult to break apart.

Nature uses this principle in our bodies to make the collagen that is the main structural component of tendons, hair, and hide. The basic unit of collagen consists of three linear protein molecules wrapped round each other in a triple helix, like a rope. Many of these triple helices are regularly packed together in bundles to form collagen fibrils. These fibrils are immensely tough and resilient; they are also highly insoluble in water. This is obvious to anyone who has tried to eat a very tough and gristly piece of meat that has not been cooked for long enough. Yet the protein molecules from which the fibrils are made are synthesised in solution—how is it that we can put together molecules that individually are soluble in a final assembly that is both highly regular and highly insoluble? If we simply were to arrange for the bonds between the molecules to be very strong, then it would be impossible to make a regular structure. Remember from our discussions of how self-assembly works, that we rely on Brownian motion to shake the structure into place, and this relies on the bonds between the molecules being strong, but not too strong. If the forces holding together the individual molecules in the triple-helix structure when the structure was originally formed were too strong, then as soon as two protein strands came into contact they would stick, whether or not they were in the perfect arrangement.

The solution to this problem is to make the structure in two stages. We start out by making a precursor molecule. This consists of the collagen molecule, with extra strands on either end. The role of these extra strands is to increase the solubility of individual strands in water. These precursor proteins self-assemble into a triple-helix structure; the presence of the soluble tails on each molecule helps this process by reducing the force driving them to associate. The presence of the soluble tail groups also keeps the triple helix in a soluble form, and it is in this condition that the collagen precursor is exported from the cell in which it has been made to the space outside the cells, where the collagen fibrils need to be assembled. Only when the collagen precursor has been transported outside the cell are the soluble tails chemically removed. At this stage the collagen molecules become highly insoluble and they pack together to form the collagen fibrils. These fibrils are then made even tougher by chemical bonds being created between different collagen molecules.

This general approach also finds application in synthetic chemistry. The conducting polymer polyacetylene is one good example. Shirikawa, Heeger, and McDiarmid won a Nobel prize for discovering that polyacetylene, unlike most polymers which are very good insulators, is a good conductor of electricity. But in the form in which it was discovered the material was practically useless; it is insoluble in any solvent and so it cannot be processed into any sort of useable form, like a film or a wire. The solution was found by the Durham chemist Jim Feast, and again it involves making a soluble precursor. Rather than directly synthesising polyacetylene, one starts out by synthesising a polymer with easily removable groups attached to the side. These groups have the effect of making the polymer soluble, so one can process the material to make a film, or a fibre, or a wire; only then, after processing, can the groups be removed to leave the conducting polymer in the desired form.

The same idea of post-processing can be taken further in the idea of templating. Once again, the idea here is to extend the use of self-assembly to hard materials. In this approach one uses a self-assembling soft system to define a template or scaffold on which a harder structure is erected. One way in which you might make this work, for example, would be to take a concentrated soap solution of the kind we discussed earlier, in which the soap molecules assembled into micelles that packed in some ordered way. Now we have a well-ordered structured that has self-assembled. Of course, the structure we have is soft. But if we have some way of petrifying part of the structure—if the water surrounding the soap micelles contains a solution that will react to form a hard mineral, for example—then we can reproduce that structure in a more permanent form. Here the soap micelles act as the template on which we build our hard nanostructure. When the hard structure has set we can simply wash away the soap.

This method is used synthetically to make materials called zeolites. The useful properties of zeolites depend on being able to control the size of the pores—as these pores are essentially the shadows of the departed soap micelles, we are using self-assembly to achieve this control even though the final material is hard rather than soft. Nature uses templating extensively—pretty much any hard structure that can be found in nature will have been made this way. Bones, teeth, shells, and the marvellously intricate exoskeletons of single-celled organisms called diatoms—all result from building a hard structure around a soft template. The properties of these nanostructured materials are quite different from the properties either of the soft template or of the hard but brittle material that is templated; they are tough, resilient, and strong in ways which are hard to reproduce, unless we try to use the same tricks.

How molecules evolve

The sheer variety of the life that exists now and has existed in the past is astonishing. The explanation for this variety, and the way in which organisms seem so well adapted to so many different ways of making a living, is Darwin's truly revolutionary notion of evolution. With this concept, we can finally understand how it is that the very different forms of the elephant and the whale, the flea and the mushroom, the grass and the amoeba, have all developed from a common ancestor.

But as we change perspective from the organism to the cell, and from the cell to the molecule, we find less and less variety in life, not more. The nanoscale soft machines on which life depends are remarkably similar, and the more fundamental they are to the business of living the more similar they seem to be. The ribosome, the machine that puts proteins together, and ATP-synthase, the fundamental machine of energy conversion—these are recognisably the same in the simplest bacteria as in ourselves. Does this mean that evolution as a concept is only important when we look at organisms? Does it mean that, at the level of the nanoscale soft machines that populate the engine house of life, evolution has little relevance? Quite the opposite; evolution is crucial to our understanding of how these machines are developed. But what we need to understand is, not how organisms can evolve, but how individual molecules themselves can evolve.

Darwin knew nothing about molecular biology, but his notion of natural selection turns out to be exactly what we need to understand this. The problem with the field of molecular evolution is that we have no molecular palaeontology, and not much in the way of comparative molecular biology. It is as if we only had a single specimen of a single species—one horse, for example—and we had to try and reconstruct the whole history of life from that one specimen, with no fossil record and no other organisms to compare it with.

This is pretty much the situation we have in trying to understand how ATP-synthase came into existence. For all their external variety, this machine is pretty much identical in every living organism. If it evolved from something simpler, then we have no trace of what it was. For all that we think that bacteria, for example, are immeasurably simpler organisms than us, at the molecular level they are fantastically sophisticated and highly evolved. If we are to believe in an evolutionary origin of life, then we have to accept that there were billions of years in which molecular soft machines were evolving from unknown primitive forms to their current state of perfection, a process of evolution that has left no trace whatsoever. We have, not so much a missing link, as a completely missing history.

But in understanding molecular evolution we have two tools that are not available in the much greater complexity of the evolution of organisms—the availability of reasonably realistic computer models, backed up by plausible physical models, and the possibility of doing experiments on a large scale.[10] It is from a combination of theory and computer simulation that we can see the need for molecular evolution in the problem of protein folding.

Remember that a protein molecule consists of a precisely-defined sequence of amino acids, each segment chosen from one of twenty. Imagine a typical protein molecule, with 100 amino acid units. How many possible sequences are there? The answer is simple enough to work out—it is 20 raised to the power of 100. This seems an innocuous enough number, until you realise that if we assembled one molecule with each sequence then the total mass of our collection of molecules would exceed the total mass of the universe. On the face of it, this would seem to render the probability of ever finding a useful protein, let alone an assembly of proteins that fit exquisitely well together and perform a beautifully engineered function, like ATP-synthase, vanishingly small.

---

[10] The difference being that you can experiment on a lot more molecules at once than even the smallest organisms. A test-tube of solution can contain many billions of billions of molecules, but a culture plate might have just one billion bacteria on it, and a laboratory's complement of house-flies might be numbered in the thousands.

At this point, we need to refer to the evolutionary theorist's favourite literary figure, Jorge Luis Borges, who wrote a famous short story called 'The library of Babel', about a library which contains all of the possible books that could ever be written. Stacked on its seemingly endless shelves are books made up of every possible sequence of letters. Every classic book that has ever been written or ever will be written is there, but the problem is that the number of even readable books, let alone classic novels, is so outweighed by the books of random gibberish that the inhabitants of the library despair of finding even a few sentences that make sense.

Our collection of possible proteins corresponds to the library of very short books, each with a hundred letters. Perhaps we should think of it as the library of all possible short poems (for example, the first verse of Blake's poem 'The tyger' has ninety-eight characters). But even this library is unimaginably huge. Suppose that we want to search for something meaningful in it. There are three levels at which we might look, according to how ambitious we are.

Somewhere hidden in the library there is the first verse of 'The tyger', and we could look for that. Lowering our sights a little, we might be happy with any poem about an animal. After a long time looking, we may well end up being happy to find a poem that simply makes sense, that is constructed from actual words put together grammatically, no matter what it is about.

Looking for one particular protein sequence in the collection of all protein sequences is like trying to find 'The tyger' in the random library; the vastness of the library makes this impossible in the finite universe we inhabit. Trying to find a protein that is effective at some task is similar to finding an animal poem. Many different protein sequences will produce a molecule that does a particular job more or less well, but as a fraction of the total number of sequences available this is still vanishingly small.

The protein equivalent of a poem that makes sense is probably a protein that folds to a single three-dimensional structure. As we saw earlier, computer simulations have shown that the property of being able to fold up into a single, well-defined, three-dimensional sequence is rather rare; from a population of random sequences, only a few sequences will be able to do this.

The essence of the idea of evolution is that it gives us a way of searching the space of all possible protein sequences in a way that is directed rather than random. We do not straight away have to find the perfect poem; instead it is sufficient to find a sequence with a word or two that makes sense, and from that we can change letters here and there, keeping any change that improves the meaning, until we have a workmanlike poem.

In the case of a protein, the first obstacle to get over is finding a sequence that reliably-folds at all. Computer simulations have shown, at least in simple models of proteins, that this property of foldability is one that we can make molecules evolve toward. Given a reliably-folded shape, then evolution will really take hold.

The first role of a protein is to catalyse a chemical reaction. The way this works is that some part of the folded protein's surface has the right shape and pattern of sticky patches that the molecules involved in the reaction are held in place in that spot in a way that helps the reaction along. Protein sequences that do not reliably fold will catalyse a reaction rather weakly, because in their collapsed state they have many distinct three-dimensional arrangements, only one or two of which might have any significant efficacy at catalysis. But as soon as the molecules learn to fold to a single three-dimensional shape, then there will be a huge jump in the potential efficiency of catalysis, and we can expect a relatively fast evolution to sequences not too different to the modern forms of the protein.

What we know about the evolution of proteins comes entirely from computer simulations. But there are some remarkable experiments that show molecular evolution taking place in the real chemists' world of the test-tube. The molecules that do the evolution are RNA. The principle was demonstrated first by Sol Spiegelman. He began with a (protein-based) enzyme—a replicase—that, in the presence of a 'primer' RNA molecule and a supply of the nucleotides that are the basic units of RNA, helps synthesise many more RNA molecules with the same sequence as the primer. He started out by mixing the replicase and the nucleotides with a primer of known sequence. After allowing the reaction to go on for some time, he would remove a small sample of the multiplied RNA and use this as the primer for another reaction. After a number of this kind of cycle, he found that the sequence of the RNA molecules had dramatically changed—the molecules had evolved.

Evolution needs some kind of selection pressure—some kind of way of deciding which of the many random changes in the molecular sequence should survive and prosper. In Spiegelman's experiment, the sequences that are preferentially selected are simply the ones that reproduce the fastest. But we can devise experiments that drive molecules to evolve more complex properties than this ability to reproduce fast in an artificial environment. If we can select out those molecules that are most efficient at catalysing some reaction, then, starting from completely random sequences, we can generate molecules that fold up in such a way that they are most efficient at catalysing that particular reaction. These molecular evolution experiments are exciting because they start to give us some clues about that distant time when life was just beginning on the Earth, and about the possibilities of life, possibly quite different in its chemistry, starting on other worlds. But they also give us some practical tips about how we can move forward from the very crude examples of self-assembly that our synthetic chemistry can provide, toward the exquisitely subtle nano-machines of nature. So far, these methods have hardly yet been exploited.

Copying nature

A factoid that is sometimes traded by nanotechnologists wishing to be controversial is that more nanotechnology patents are held by the cosmetics company

L'Oreal than either IBM or Intel. One can debate the significance or relevance of this,[11] but it does emphasise one underappreciated truth, which is that progress in nanotechnology may come from unexpected sources, and businesses that make their money from detergents and block copolymers are well placed to appreciate how self-assembly can be used to make artificial nanostructures.

What, then, can we learn from studying the way that nature puts together soft machines? Can we use some of these principles to take synthetic self-assembly beyond making shoe soles or curiously-textured shampoos? It is in the area of soaps and surfactants that the most progress has been made. One of the most fruitful concepts has been the idea of a liposome (or a vesicle), a capsule made from a lipid bilayer that has folded over itself to create an enclosure, like an artificial cell wall. Liposomes, at times, feature strongly in the advertising of particularly expensive lotions for supposedly rejuvenating the skin, but their prominence in the dubious trade of advertising does not mean they are not genuinely important as a potential means of encapsulating and delivering some active ingredient. Liposomes are relatively easily made from quite simple ingredients, and they have been extensively explored as vehicles for delivering drugs.

Surfactant vesicles also play a major role in artificial attempts to exploit the idea of templating and scaffolds. The idea here is to use vesicles to create nano-sized particles of some hard material. For example, one might want to make tiny particles of a semiconductor, whose optoelectronic properties are modified by quantum effects due to their size. If we can make a compound AB by reacting substance A with substance B, then our strategy would be to create vesicles which contain a solution A, and then to expose the solution of vesicles to a solution of B, relying on the ability of B to pass through the thin and fluid wall of the vesicle, to create a particle of our target compound whose size is defined by the sizes of our original vesicles. This strategy has been successfully employed to make all kinds of nano-sized particles, which have already found a variety of practical uses.

The templating strategy can be extended to reproduce in hard materials all of the remarkable nano-sized shapes and topologies that can be simply created by self-assembly in soap solutions. What about the more specific and carefully-engineered types of self-assembly used by nucleic acids and proteins? The perfection of the base-pairing mechanism has led a number of researchers to try to exploit DNA to make self-assembled structures of remarkable complexity.

Few have been more imaginative than the pioneer of this field, Nadrian Seeman. He realised that you could design sequences of DNA, so that, for example, three strands would come together to make three segments of double helix that came together at a single junction. From this kind of basic unit you could go on to design sequences that created all sorts of patterns: knots,

[11] It was certainly true in 2003 that if you searched the US Patent Office database for patents with the element 'nano' in the title or abstract, then there were many more assigned to L'Oreal than to IBM or Intel. This, of course, must also reflect the different ways in which scientists from different organisations like to label their work.

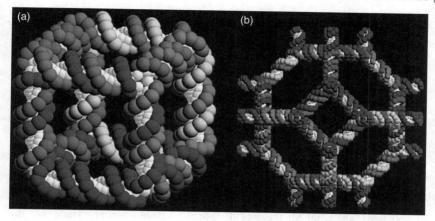

Fig. 5.9 Complex structures made by the self-assembly of DNA molecules with designed sequences. Reproduced with permission from N. C. Seeman. Biochemistry and Structural DNA Nanotechnology: An Evolving Symbiotic Relationship *Biochemistry* **42** (2003) 7259–69. Copyright 1996 American Chemical Society.

ladders, and three-dimensional structures like wire-frame representations of cubes and octahedrons (see figure 5.9). Seeman was using the base-pairing mechanism to achieve the same goal of specifying three-dimensional structure by one-dimensional sequence that nature achieves, but by a slightly different detailed mechanism.

Seeman's visions are still not at the point of being close to practical applications. I first came across Seeman in the incongruous surroundings of a country house hotel on the edge of Dartmoor, where a mixture of industrial and academic polymer chemists have an annual conference and a chance to get away from their less salubrious normal working places in chemical industry centres like Middlesbrough and Grangemouth. Seeman gave a great lecture, illustrated with intricate computer graphics of ladders, cubes, and knots made out of DNA; and then, at the end, someone asked him how much of the material he actually made in an experiment. 'Why, about 10^9 molecules', he replied. I watched first the concentration on the faces of the audience members as they did the mental arithmetic needed to work out what mass this corresponded to, and then the incredulity when they realised just how small it is—a tiny speck at one-billionth of a gram, and this to a profession that would regard 10 kg of material as a small experimental batch.

But Seeman had the vision to realise that molecular biology is being quickly industrialised; now adverts in the journal *Nature* compete to offer custom synthesis of DNA for as little as 80p per base. You simply send them the sequence you want and the money, and a couple of weeks later a little trace of DNA comes back. At the laboratory level, we have at least one demonstration that we can use the full power of self-assembly to make, not a single soft machine, but billions of them all at once.

6
Machines and mechanisms

Introduction

'I sell here, sir, what all the world desires to have—power.' These are the reported words of Matthew Boulton, the early industrialist, who collaborated with James Watt to manufacture a steam engine of hugely improved design. It was this device, which, perhaps more than any other, made possible the industrial revolution. Before then, if you wanted more work done than human muscle power could manage, then you could use animals like horses or oxen, you could use a windmill, or you could use water power. Animals need to rest, with wind power you are at the mercy of the weather, and the use of water power on an industrial scale relies on a suitable combination of climate and topography. England's Peak District, where I am writing these words, combines a hilly landscape with a quite stupendous propensity for rain, and, as a consequence, was industrialised very early on the basis of its water power, but the inhabitants of drier and flatter places were less lucky. So the development of power sources which had much greater capacity than animals, which were reliable, and which did not depend on a favourable location, provided both power and wealth to their inventors. At the beginning of the industrial revolution, the ultimate power source was the energy won by burning coal. The energy in coal is chemical energy; this chemical energy is converted into heat energy by burning it in air, and the heat energy is converted to mechanical energy by boiling water to drive a steam engine.

What is going to drive an industrial revolution based on nanotechnology? One thing we can be sure about is that it will not be steam engines, or indeed any of the wide class of engines scientists refer to as heat engines. A heat engine relies on the ability to keep one part of the engine at a different temperature to another part, and in very small systems this becomes increasingly difficult. Instead, we need to have an engine that converts chemical energy directly into mechanical energy, without, at any time, raising the temperature of the surroundings. We come across very few examples of this kind of constant-temperature motor in our everyday lives—in fact, the only one I can think of is the old-fashioned toy in which a piece of camphor is used to propel a boat across the surface of one's bath—but biology relies on them. Very early in the evolution of life, motors based on single molecules were developed to convert chemical energy into mechanical energy in the constant-temperature conditions that are appropriate to the nanoworld. Many simple bacteria have molecular motors that they use to propel themselves around. In more complex, but still single-celled, organisms, molecular motors transport goods from place to place in the cell like little trains running on tracks. In animals such as ourselves, many highly-specialised cells are combined to make muscles; the substantial mechanical power that a big animal like a horse can develop is simply the result of adding up the nanoscopic power outputs of vast numbers of molecular motors working together.

When we think of engines, the quality that we imagine that their components must above all possess is rigidity. What is so striking when you look at, say, a large working steam engine, is the mass and solidity of the steel pistons, heavy flywheels, and giant beams. The idea of making an engine from soft stuff, from rubber or jelly, seems quite absurd. But it is this kind of soft matter that biology is made from. A slab of meat is tender and soft. But when it was in a living animal it was an engine, able to generate a degree of power that, weight for weight, is not much less than a petrol engine.

You might think that it is a remarkable testament to the ingenuity of nature that, through evolution, it developed such effective ways of using such unpromising materials. But this would be getting things exactly the wrong way round. The softness and floppiness, which at the macroscopic scale seems so ludicrously inappropriate for making mechanical devices, is, at the nanoscale, exactly right. It is not that biological molecular motors work in spite of the floppiness of their components; they work because of it. The operating principles of biological molecular motors exploit the lack of rigidity of biological molecules. It is this lack of rigidity, together with the ceaseless pummelling of the molecules by the randomly-moving water molecules, that leads to Brownian motion, which underlies the changes in molecular shape that are the ultimate causes of the power strokes in biological engines.

When we come to try to make artificial nanoscale machines, we have a choice between two design philosophies. On the one hand, we could use the same engineering design principles that have served us so well on the macroscopic scale. These rely on rigid materials, components that are fabricated to precise tolerances, and the mutually free motion of parts with respect to each other. As we attempt to make smaller and smaller mechanisms, the special physics of the nanoworld—the constant shaking of Brownian motion and the universal stickiness that arises from the strength of surface forces—will present larger and larger obstacles that we will have to design around. On the other hand, we could use the design principles used by cell biology—what we might call 'soft engineering'. These are very unfamiliar, even counter-intuitive, because they exploit phenomena, such as molecular shape change, that have few counterparts at the macroscopic level. But the advantage of soft engineering is that it does not treat the special features of the nanoworld as problems to be overcome, instead it exploits them and indeed relies on them to work at all.

Prime movers—engines large and small

'Heat is work and work is heat', as Flanders and Swann chanted in their sung version of the laws of thermodynamics. The first of these laws tells us that there are many different forms of energy, but these forms are interchangeable, with the condition that the total amount of energy is always conserved. The operations of industry require that we do useful work—we move and lift objects, we position things, we cut them, and join them. The energy required to do this work needs to be obtained from some other source, and converted to work by some kind of engine. We ourselves convert chemical energy, stored in the food we eat, into mechanical energy by the operation of our muscles. As our material culture developed, manpower was successively replaced, first by the power of larger animals, and then by engines which converted the energy of wind and water into useful work. The industrial revolution was made possible by the development of engines that convert the chemical energy stored in fossil fuels into useful work.

For much of the nineteenth century, the dominant prime mover was the steam engine, in which burning coal vapourises water to produce steam. The pressure changes that occur when water is boiled and steam is condensed are exploited to create mechanical work, to drive machinery or propel a locomotive. The source of energy in a steam engine is the burning of a fuel, but, as this takes place outside the engine proper, the steam engine is referred to as an external combustion engine. The discovery of oil and the techniques to refine it led to the possibility of the internal combustion engine, in which the fuel is burned within the engine itself, directly creating the increase in pressure within a cylinder that forces the movement of a piston. Petrol engines and diesel engines are light, portable, and convenient, and made possible the development of the automobile and the truck. In another development toward the end of the nineteenth century, it was realised that high-pressure steam could be used to cause a rotor with a ring of blades to rotate within a fixed casing. These steam turbines generated rotary motion directly, and were ideal both for turning the screw propellers of ships and for driving electricity generators. When we operate an electrical appliance we are using an external combustion engine at one remove; energy is converted from chemical energy to heat energy at the power station, where coal, oil, or gas is burnt to create high-pressure steam. The heat energy is converted to mechanical energy in the turbine, this in turn is converted to electrical energy in the generators, and, after transmission to a house or factory, the electrical energy can be converted back to mechanical energy in an electric motor. A variation on this comes when we store the electricity in rechargeable batteries—here we convert the electrical energy to chemical energy, later converting the chemical energy back to electrical energy and then mechanical energy. So, even our electric toothbrush is ultimately powered by steam.

Our comfortable, industrial-age life is made possible by the conversion of chemical energy to mechanical energy through the intermediary of heat. Steam engines, petrol engines, diesel engines, gas turbines, and jet engines are all examples of the generic class of engines called heat engines. A heat engine works by exploiting a temperature difference. As heat flows from where it is generated to the exhaust, some of this energy can be captured and converted into work. In a steam engine, the source of the energy is the coal fire that heats the boiler; in a petrol engine, it is the explosion of the petrol/air mixture inside the cylinder. At the macroscopic scale, heat engines so dominate our lives that it is difficult to imagine any other way of converting chemical energy to mechanical energy. But when we try to make things on a nanoscale, we will no longer be able to rely on heat engines as prime movers.

For a heat engine to work well, there needs to be a big difference in the temperature of the hot end of the engine and the cold end. In fact, it is a fundamental result of thermodynamics that any heat engine cannot exceed a maximum efficiency that is simply determined by the temperatures of the hot and cold reservoirs. To be precise, that maximum efficiency is found by dividing the difference in temperatures between the hot and cold ends of the engine by the absolute temperature of the environment.[12] For example, a power station will use a steam turbine to convert the heat energy derived from burning coal or gas (or from a nuclear fission reaction) into electrical energy. A steam turbine might be run at a temperature of 540 °C, which gives a maximum thermal efficiency of around 60%. This is a theoretical upper limit, and the practical limit will be a bit less than this. The maximum thermal efficiency can be increased by increasing the operating temperature, and the introduction of new alloys for turbine blades in recent years has allowed significant increases in efficiency.

Almost all students of physics, chemistry, and engineering find this classical thermodynamics a fantastically boring subject, an impression that is not helped by the quite correct perception that it was all worked out in the nineteenth century so that better steam engines could be made. What has such a Victorian branch of physics got to do with modern science, let alone nanotechnology?

The answer lies in the familiar subject of how quickly things go cold. Where I live, cafés often offer you the choice of having your cup of tea in a big mug that holds one pint, or a smaller half-pint mug (a delicate quarter-pint cup of the kind favoured by the more refined establishments would be regarded as a bit of a waste of time). If you are served a pint mug and a half-pint mug at the same time, then you have to wait appreciably longer for the larger mug to get cool enough to drink. Small things lose heat more quickly than big things. If the dominant mechanism by which the heat is lost is by the thermal conductivity of the material (and this mechanism will tend to dominate on small scales), then the rate of loss of heat will be proportional to the square of the size of the object. So, if my pint mug of tea takes half an hour to

---

[12] The absolute temperature is the temperature measured relative to absolute zero, at which all thermal motion stops. Absolute temperature in the Kelvin scale is obtained from the temperature in Celsius by adding 273.15.

become too cold to drink with pleasure, then the heroes of 'Fantastic voyage' would have to be very prompt in drinking their one-micron mugs of tea, because they would be cold in a fraction of a microsecond.

As we try to make a heat engine smaller and smaller, we will need to run the engine faster and faster in order to get some work out of it before the heat in the working fluid—the steam in a steam engine, or the hot gases in the cylinders of a petrol engine—leaks out through the walls of the engine. But the decreasing size of the engine also puts limits on how quickly we can make it run. Even for a piston made of the most rigid materials, if we push one end of the piston rod then it takes a certain finite time for the other end of the piston to respond by itself moving. If we try to move the piston back and forth at a rate that is greater than the rate at which the piston rod can respond, we will not be able to extract any work at all from our engine.

So it is going to be difficult to maintain any significant differences in temperature in the nanoscale world. The ultimate cause of heat conduction is, once again, Brownian motion, the random motion of molecules that so dominates the nanoscale world. In a liquid, molecules in a local hot spot are moving fast, physically taking their energy away from their starting-point and passing it on to neighbouring molecules by colliding with them. In solids, atoms themselves cannot move significant distances, because they are bound in their well-defined lattice sites in the crystalline structure of the material. But they can vibrate, and organised patterns of vibrations of a number of atoms[13] behave as little packets of heat energy, spreading that energy throughout the material. So it is the restless, random motion that has been seen to dominate the world of the nanoscale that renders our macroscopic designs for engines unworkable. To make a workable nanoscale motor, our best bet is to find a mechanism that exploits this random motion, but to do this we need to understand another law of thermodynamics—the second law.

Maxwell's demon

A glass of water looks like a fairly placid object, but we now know that if we could look at it a million times magnified then we would see constant activity. The molecules that the water is made out of are moving around at high speed, constantly colliding and changing direction. Any object suspended in the water—a dust particle, maybe, or a microbe that has escaped the purification system—would be constantly bombarded from all sides by the fast-moving water molecules, and as a result of this battering it will jiggle around restlessly. It is a high-energy environment, the energy simply arising from the heat that the water contains.

Is it possible to take this random, purposeless, and undirected motion, and do something useful with it? Could we, for example, extract some of the

---

[13] Known in solid-state physics as phonons.

energy and use it to do useful work? Unfortunately, the answer to this question is no; it is forbidden by one of the fundamental laws of nature, the second law of thermodynamics. But even though we cannot directly extract energy from the thermal motion of molecules and nanoscale objects, we can use their motion to drive mechanisms that convert energy on the nanoscale. A nanoscale motor can be built which effectively harnesses Brownian motion and puts it to useful work, though it must have an energy input to do this.

It is a pity that we cannot extract the heat energy directly from a tank of water and use it to do useful work, because if we could, this would completely solve all of our energy problems. The impossibility of doing this has not stopped people trying, though. A history of perpetual-motion machines would describe an alliance of untrained visionaries and out-and-out hucksters who have proposed a series of machines that purport to do just that. One of the earliest of these was an inventor called John Gamgee, who in 1880 persuaded the US Navy that you could use the heat from the ocean to boil liquid ammonia, the expansion of which would power your warships. The ammonia would then be condensed to begin the cycle again. Of course, in reality you would need to cool the ammonia down to get it to liquefy again, and this would use up more energy than you got from the engine. This fundamental flaw did not stop later inventors from rediscovering this kind of motor, and, even today, foolish people in the United States who did not pay enough attention to their high school physics lessons are being parted with their savings in scams based on this idea.[14]

Viewed from a macroscopic point of view, the reason why useful energy cannot be extracted from our surroundings is that this would counteract the second law of thermodynamics. This can be formulated in many ways, but the one most directly relevant to this discussion is the common-sense observation that heat only flows in one direction, from hot to cold. To extract energy from something like a tank of water, which has a constant temperature everywhere, we would need one part of the tank to spontaneously get colder, as the energy from this cold spot flowed out into our engine. But we know, from long experience, that a warm gin and tonic sitting in a warm room does not spontaneously grow ice-cubes, and cold cups of tea do not heat themselves up again by drawing in heat from the room.

At a microscopic level, what we would need to do to get such a machine to work is to take the completely disordered, random motion of the molecules, and put some order into it, to give the molecules a bit of discipline. But molecules will not discipline themselves, and the energy you need to put in to discipline them will always outweigh the returns. This is best shown by consideration of a thought experiment proposed by the nineteenth-century physicist, James Clerk Maxwell. He imagined the operation of a device, which would sort out fast molecules from slow ones, and thus solve the problem of extracting useful energy from heat.

[14] The recent history of the perpetual-motion machine is entertainingly described in Robert Park's book *Voodoo science* (Oxford University Press, 2000).

This non-existent device is now universally known as Maxwell's demon. Imagine two containers, each filled with a gas at room temperature. Between them is a door. If the door is always open, then the two containers will stay at the same temperature. The molecules are moving around at high speed, colliding with each other, and colliding with the walls. At any time, some of the molecules will be going faster, some slower. Some fast ones will go through the door from left to right, increasing the average energy on the right-hand side, but almost immediately these will be balanced by some traffic in the opposite direction.

Now our demon comes on duty. His job is to guard the door; if he sees a fast molecule coming from left to right, then he opens it. But he keeps the door barred for slow molecules. The result is that the average energy of the right-hand side gets higher, and the average energy of the left-hand side gets smaller. We have started out with a gas all at one temperature, and we have managed to end up with one part of the apparatus heating up, and one part cooling down.

How might such a demon work? At its simplest, we might imagine a door which could only open in one direction, which was kept open by a spring. The strength of the spring was such that only the fast molecules could open it, while the slow ones would bounce back. Why would not this kind of spring-loaded door-work? The answer is in the universality of Brownian motion. As we saw in Chapter 4, everything that is at a finite temperature undergoes Brownian motion. Even without the collisions from the gas molecules, the door would be jiggling around, opening and shutting at random. It turns out that this random opening and shutting will always prevent the door from doing its job of sorting the quick from the slow. Moreover, no matter how complicated the mechanism of the demon, Brownian motion will always defeat its efforts to sort the hot from the cold.

Isothermal motors

We cannot directly extract any of the energy of the randomly-moving molecules in a liquid, because the second law of thermodynamics keeps this energy firmly locked up. At the nanoscale, we cannot make a motor that relies on sustaining a temperature difference, because the random motion of molecules will always, if our system is small enough, even out the distribution of heat energy before we have a chance to convert any of it into work. How can we make a motor that works in conditions of constant temperature?

One approach to making a motor that operates at constant temperature—an isothermal motor—is to recognise that, even though we cannot keep heat confined to one place at the nanoscale, we can keep molecules confined. Just as a difference in temperatures between two locations can drive an engine, so can a difference in the concentration of molecules. The camphor boat provides a nice illustration of this.

This toy consists of a little boat that can be powered across the surface of a bath at quite surprising speeds. It is driven by a little chip of camphor (an organic solid with a distinctive medicinal smell, that used to be common in households in the form of moth-balls) lodged on the water-line at the rear of the boat. What drives the boat forward is the fact that, as the camphor dissolves from the chip, not all of the molecules go into solution; instead a layer of molecules forms at the surface of the water. These molecules are moving around in all directions; buffeted by collisions with fast-moving water molecules, they undergo Brownian motion in two dimensions. Fast-moving camphor molecules collide with the boat, each time giving it a little kick. But because it takes time for the molecules to spread uniformly across the surface, the back of the boat, near the source of the camphor, receives more kicks than the front and there is a net force propelling the boat forward.

Understanding the overall flows of energy that drive the camphor boat is a little more subtle than for a steam engine. At the beginning of the process, the camphor molecules are packed closely together in the crystal, glued together by relatively weak intermolecular forces. At the end of the process, the molecules are rather uniformly dispersed throughout the surface and bulk of the water. What drives this process of dispersion is the universal tendency towards states of increasing disorder that is mandated by the second law of thermodynamics. We will see other examples of this later in this chapter.

In the example of a camphor boat, we can see a very concrete example of the way in which Brownian motion can be used to drive an engine, even in the absence of any gradients in temperature. Jacques Prost proposed a very general mechanism by which Brownian motion can be rectified to extract useful work from the random motion of molecules—a model known as the Brownian ratchet (see figure 6.1). The Brownian ratchet is a rather abstract model in which we imagine a potential which has the shape of a saw-blade, with one steep side and one gentle side on either side of each point. This potential is periodically turned on and off. Imagine releasing a set of molecules at one place. If the potential is not there, then the molecules simply diffuse away by Brownian motion. Their average distance from the starting-point increases with time, but since they are equally likely to go in any direction there is no overall useful motion. Now we imagine turning the potential on. The molecules run downhill into the bottom of the valley that they find themselves in. But because the valleys are not symmetric in shape, there is a tendency for molecules to move, on average, more in one direction than the other. Despite the fact that their motion, driven as it is by Brownian motion, is entirely random in direction, we end up with a net movement of molecules in one particular direction.

Prost's Brownian ratchet is typical of a physicist's model, in that it is highly abstract, and its purpose is less to allow one to design a working motor that exploits Brownian motion in a system at constant temperature, and more to set one to thinking about the general principles that are at stake if one sets oneself this task. We will now move on to a concrete and everyday problem—how an

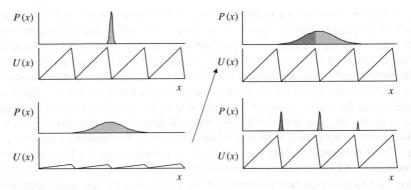

Fig. 6.1 The Brownian ratchet. Here $P(x)$ gives the probability distribution for a number of particles that are randomly moving by Brownian motion; $U(x)$ is an asymmtric saw-tooth potential that is alternately turned on and off. At the beginning of the cycle, the particles are stuck at the bottom of one of the potential wells. Then the potential is turned off, allowing particles to randomly diffuse away from their starting position. When the potential is turned back on, the particles move downhill, falling into one of the potential wells. Because of the asymmetry of the potential, more particles fall into the well to the left of the starting well than the well to the right. There is a net flow of particles to the left, even though the random motion of the particles has no preference for left over right.

elastic band works—which gives us a surprising insight into how we might in practise exploit entropy and Brownian motion to do useful work.

Rubber elasticity and the entropic spring

Even if we cannot extract the energy from the Brownian motion of a material which is all at one temperature, there are circumstances in which we can harness this random motion to make something more directed. To do this, we need to start out from a special state, one in which we have artificially increased the degree of order in one part of the system. As we discussed in the previous chapter, and as anyone who has a small child knows, order is not the natural condition of the universe, and as soon as one has tidied something up the forces of disorder soon untidy it. Or, to be more scientific, if one prepares a system in a state of lower-than-maximum entropy, then the operation of the second law of thermodynamics ensures that the system will evolve into a state of higher entropy. The mechanism by which the entropy will be increased is by the collisions with the randomly-moving particles around it—Brownian motion, in other words. If one can use this tendency to achieve a state of higher disorder to do some useful work, what one has in effect done is converted some of the random, Brownian motion of the environment into useful, directed effort.

This is rather an abstract discussion, but there is one very important application of it that will be familiar to everybody, and that is a rubber band.

If you stretch a rubber band, then you feel a force, and if you are, for example, a small boy, then you can use this force to your own ends, for example, by using a catapult to fire a missile. When you make a catapult by stretching a rubber band across a piece of forked stick, and you use it to fire a projectile, where has the kinetic energy that projects the speeding projectile come from? Ultimately, from your own muscles as you pull the elastic back, storing the energy in the elastic energy of the elastic band. But as the band is released, the origin of the force that contracts it and propels the missile onwards is actually the Brownian motion of individual segments of the rubber molecules.

To see how this works, we must first remind ourselves of what a polymer molecule looks like. Rubber comes from the sap of a tropical tree,[15] but after it has been harvested it is very similar in nature to plastics such as polythene and polystyrene. All of these materials are polymers; a polymer molecule consists of many identical chemical units joined together to make a long chain. The links of the chain are rather flexible, so if one imagines a single molecule immersed in a liquid then one should imagine the links of the chain being continuously bombarded by the ever-present Brownian motion of the solvent molecules. In response to this bombardment, the chain will be forever flexing and folding. The shape that the chain makes in space will be highly disordered and highly folded back on itself. In fact, what determines the shape will be, yet again, the tendency things have to maximise their disorder, or entropy. The most ordered shape that the chain could have would be to be entirely stretched out, as a long, straight rod. We can quantify this degree of ordering; there is only one way to arrange the molecule so that it is fully stretched, but there are many ways of arranging it so that it is crumpled up.

If we could take a polymer molecule, and stretch it out, then we would be forcing it into a state which is more ordered than it would like. The tendency that every system has to go from a state of high order to a more disordered state will actually give rise to a physical force pulling the two ends together. It is this entropic force which leads to the elasticity of rubber.

The idea of this rather abstract sounding 'tendency to disorder' resulting in an actual physical force sounds rather unlikely—what is actually pulling the two ends of our polymer molecule together? It must be the Brownian motion of the molecules surrounding it. Each segment of the polymer, each link of the chain, is being pummelled by fast-moving molecules colliding into it, and in response to each of these blows the molecule will fold up. If the molecule starts out straight, then each blow that falls on the molecule will make it more folded, pulling the ends in together, until after many such blows the molecule is so folded up on itself that a punch is as likely to straighten that bit of the molecule as to fold it up further. When this state has been reached the molecule is in its state of maximum entropy, a relaxed condition in which there is, on average, no force pulling the ends together.

---

[15] Synthetic rubbers are also available, but the natural stuff is very widely used and for many applications is still better than the artificial substitutes.

A rubber band is slightly more complicated than this, because we do not simply have a single polymer chain. In making a rubber band from the sap that comes from a tree, there is one more essential process. Unprocessed rubber is rather soft and quickly wears out. It is made up of a mess of very long polymer chains that are completely tangled up with each other. The reason that it does not behave more like a liquid is simply that the polymer chains are too knotted together to be able to flow past each other easily. Nonetheless, given enough time the knots will disentangle, helped by the constant shaking of Brownian motion, and the rubber will yield. The key to making rubber a material that is robust and resilient as well as elastic was found by Charles Goodyear in 1844. He discovered that, by heating rubber with sulphur, it became somewhat stiffer and much better able to resist flow. In this process, known as vulcanisation, what is happening is that, at various points along the length of the polymer chains, pairs of chains become chemically joined to make a network. A piece of rubber, then, is a giant three-dimensional molecule in which one could trace a continuous path, linked by permanent chemical bonds, between any two points in the material. Between the points at which chains are joined—the cross-links—the strands of polymer chain are still highly folded and coiled, and if there are not too many cross-links then they are still free to move around.[16] A rubber is a curious material, because on the macroscopic scale it is a solid, because it cannot flow, but if you look at it on a fine enough scale then it would look like a liquid, with the individual segments moving around as far as the constraints from the cross-links allow them. This means that individual segments are undergoing Brownian motion, and that, if a strand of chain is stretched out straight, then the collisions of all of the surrounding molecules cause it to crumple and fold, just like a single chain in a liquid. So the macroscopic force you feel when you stretch a rubber band is just the result of the Brownian motion of the segments that make up the rubber acting by crumpling up the strands of chain linking the cross-links.

The role of entropy in causing the elasticity in rubber has been understood for a long time, but recently the phenomenon of entropic elasticity has been dramatically illustrated experimentally on single chains. Using an atomic force microscope, it is possible to attach one end of a polymer chain to the microscope's tip, with the other one being tethered to a flat surface, and directly measure the force as the tip is moved away from the surface. The forces arising from stretching an individual molecule are tiny. It would take a few hundred billion molecules to pick up an object with the weight of an apple. But in a rubber band this is exactly what one has, many individual strands of polymer chain, all exerting a force because of the buffeting they receive from Brownian motion.

In the entropic elasticity that makes rubber bands work, we have a mechanism that converts the random collisions of Brownian motion into a useable, directable force. But this does not yet help us make a motor because the elasticity of a

---

[16] Rubber which is vulcanised to a very high degree forms a hard, glassy material known as ebonite, which, as its name suggests, used to be used as a substitute for hardwoods like ebony.

rubber only allows us to store energy. When we stretch a rubber band, we have to do work to pull the ends apart, and this work is stored in the elastic energy of the band, to be released when we let the ends go. Can we devise a way of using entropic elasticity to convert energy, to make a real motor? To do this, we have to understand how the random, crumpled shape that a flexible chain takes up when it is being continuously buffeted by Brownian motion is altered when some or all of the segments have the property of sticking to each other.

Chain collapse and responsive gels

'Solid as a rock', we say, when we want to evoke the immutability of material objects, perhaps in contrast to the changes that living things undergo. But a material object that could change its shape or form could have its uses. The idea is a very old one in literature and in science fiction—notably in films like 'Terminator', where modern computer graphics make it very realistic. The idea of self-assembly gives us a clue how to do this—if the structure that is encoded in the molecule is different, or if the environment is different, then changing the environment could cause the shape to change.

When we talked about self-assembly we imagined a three-dimensional jigsaw in the shape of a rabbit. This was made out of pieces that had weak glue on the faces that would be in contact when forming the desired rabbit shape. When we put all of the pieces in a bag and shake them, if we get the balance right between the violence of the shaking and the strength of the glue then we imagined getting a fully-formed toy rabbit by the process of self-assembly. Now imagine stretching this already unlikely scenario one step further. Suppose that the pieces have been very cunningly designed so that if they are put together one way then you get a rabbit, but if you put them together another way then you get a dinosaur. Now imagine that you have two types of glue, one of which is sticky in water, but which does not stick at all in air. The other glue is exactly the opposite; it sticks well in air, but not at all in water. Now we put the glue that works in air on the faces that go together to make a rabbit, but on the faces that are in contact for the dinosaur we put the glue that works in water. Now we can make two different toys from the same set of pieces. If we put all of the pieces in a bag in the air and give them a shake, then what comes out is a rabbit. But if we put the pieces in a bucket of water and give them a stir, then what we get is a dinosaur. We have a single set of components that will self-assemble into two completely different shapes depending on what environment was present when the self-assembly process took place. Even more significant than this is what happens if we put a fully-formed toy rabbit into a bucket of water. Now the glue that held the pieces together in the rabbit shape stops working and the rabbit falls to pieces. But the second glue now starts to work, and the pieces reassemble in the shape of a dinosaur. We have an object that can change its shape in response to a change in the environment.

In the macroworld, this is starting to look like a far-fetched idea. But the special conditions in the nanoworld are much more conducive to this property of responsiveness, this potential mutability of material objects. Brownian motion is always with us, shaking things up, all the time testing whether there is a better way of sticking pieces together. The nanoworld is not a conservative place. Surface forces are strong in the nanoworld, but it turns out that there are various relatively easy ways of changing the relative strength of different surface forces, so that interactions, which in one environment would stick things together, in another environment might be outweighed by another repulsive interaction.

How might this work in reality? The simplest example is a simple, synthetic macromolecule which is made up of many identical units strung together in a long chain. These are molecules of the sort that make up plastics like polythene and polystyrene. We rather dismissed the possibilities for these very simple molecules doing anything interesting in the self-assembly line; they are just too simple. But they do illustrate quite well the idea of responsiveness. So if we take a plastic material, and immerse it in a solvent, then it will dissolve, because each polymer segment sticks better to a solvent molecule than it would to another polymer segment. In this state, each polymer molecule takes the form of a highly-extended, fluffy ball. The individual segments of the chain are constantly in motion—Brownian motion—but if two segments should collide then they will soon come apart again, as there is nothing to stick them together. This fluffy open structure—known as a coil—will be constantly flexing and changing its shape in detail, but on average it will look pretty constant over time—with an overall size that varies as the 3/5th power of the total length of the molecule. This scaling shows how different an object this is to a conventional solid object; if we screw up a piece of string into a ball then the volume of the ball is proportional to the length of the string. Therefore the overall diameter of the ball varies as the 1/3rd power of the length of the string. This non-standard scaling of size with the total amount of material is typical of objects with the character of fractals.

The fluffy, open, fractal character of a polymer chain in a good solvent—a solvent which forms stronger interactions with the polymer segments than the segments do with each other—depends on two things—the fact that the polymer segments are in constant, random, Brownian motion, and the fact that if two segments collide then they will almost immediately come apart again, because they do not stick. But if we can change the environment so that the segments do want to stick, then we will see a very pronounced change in the shape of the molecules.

To do this, we could perhaps change the temperature, or (more simply, from an experimental point of view) add a second solvent to the mixture, which mixes well with the first solvent but which does not interact so favourably with the polymer. In this way, we move to a situation in which the polymer interacts more favourably with itself than with the solvent. Now when two polymer segments collide with each other as they are being moved around by Brownian motion, they are likely to stay stuck. If two distant parts of the chain are pinned together in this way, then the chances of other pairs of segments colliding increases, and when they collide they in turn stick together too. The result of this is a rapid collapse of what is initially a very open, fluffy structure to a compact blob of almost pure polymer—a globule. This transition between a coil and a globule happens very abruptly, and the change in size of the molecule can be very large. Roughly speaking, the factor by which the coil shrinks is given by the number of units in the polymer raised to the power of 4/15. As it is not that difficult to find a polymer made up of 10 000 segments, this means that we have a molecule that can increase or decrease its size by a factor of ten as the environment changes, going in the process from a structure that is very loose and fluffy to one that is rather compact. The process is quite reversible; if we change the solvent back to what it was, perhaps by adding more of the original good solvent or by changing the temperature back to what it was, then Brownian motion will be able to shake apart the now less-well-stuck pairs of segments, and soon the expanded, coil state will be restored.

Actually, it is quite difficult to see a simple coil–globule transition of the kind that I have just described in a practical way, for one simple and obvious reason. If we make up a solution of polymers then we have more than one molecule, and when two segments come together they behave the same way whether they belong to the same chain or to different chains. Thus, in real life, unless the concentration of polymers we use is very low, what happens is, not that individual chains collapse neatly to globules, but that many different chains stick together to form an insoluble precipitate, which falls out of the solution and forms a sticky mess at the bottom of the test-tube.

There is one way in which we can see an analogous effect at a macroscopic scale. As we discussed above, rubber is a network in which the molecules are linked together to form what is effectively a giant molecule, loosely linked together in three dimensions. If we take a piece of rubber and immerse it in a solvent that has a favourable interaction with the segments of the rubber molecules, then the network will swell, drawing in the solvent. But this process of swelling must have a limit; as the solvent is drawn in, the strands of polymer between the junction points get stretched, and Brownian motion will act to fold these strands up again. The final size of the swollen gel will depend on a balance between the tendency of Brownian motion to contract the strands of polymers, and the efforts of the solvent to make contact with as many segments as possible.

This means that if we can change the environment so that polymer segments will tend to stick to each other, rather than to the solvent molecules, then the polymer network will shrink, expelling the solvent. This is very similar to the situation for a single chain, but now the transition from an open structure to a compact structure occurs on macroscopic dimensions. We can tailor the chemistry to choose from the same range of environmental stimuli to trigger the response as drive the collapse of a single chain. Gels which respond to temperature have been most closely studied. Another very attractive possibility is to have a gel which responds to a change in the degree of acidity or alkalinity of a solution. Acid solutions are characterised by a high concentration of hydrogen ions.[17] A weak acid is a substance from which a hydrogen ion will split when it is in a solution containing few hydrogen ions (i.e. in alkaline conditions), leaving a negatively-charged residue. In an acidic solution, on the other hand, the weak acid stays intact, and because of this has no charge. A polymer with weakly-acidic groups will change from being ionised in alkaline conditions, in which state it is highly soluble in water, to being neutral in acidic conditions, in which case the polymer segments are rather hydrophobic. This kind of weakly-acidic polymer will have a transition from an open, expanded state in alkaline conditions to a collapsed, globular state in acids.

In fact, the most common everyday example of this piece of physics will be familiar to many new parents. Modern disposable nappies (diapers in the USA) rely on this kind of gel transition. The gel particles that are so efficient at absorbing baby urine are made from a cross-linked network of a weak polymer acid. Synthesised in the swollen state, they are then dehydrated to yield collapsed, dry gel particles. It seems that baby urine is just alkaline enough to make sure the polymer becomes ionised, giving the network the capacity to soak up a truly impressive quantity of baby urine.

Can there be a more glamorous application for a gel transition than nappies? The idea of a reversible transition in shape and properties driven by some external stimulus suggests all sorts of possibilities, most notably devices to deliver drugs in response to changes in temperature or other stimulus. Intriguing toys have been made that use gel collapse to make a fish swim or a worm wiggle. But serious applications have been fewer than perhaps had been anticipated when the phenomena were first discovered and explored, most notably by the late Toyoichi Tanaka at the Massachusetts Institute of Technology.

Fig. 6.2 A pH sensitive gel in its swollen (left) and collapsed (right) states. Courtesy Jon Howse, University of Sheffield.

---

[17] The chemist's measure of acidity, the pH, is simply the negative of the logarithm of the hydrogen ion concentration. The lower the pH, the higher the hydrogen ion concentration and the more acidic the solution.

This is because gel collapse is a phenomenon that works better in small things than large ones. In a large lump of gel, the rate at which the material can shrink or swell is limited by how fast the solvent can squeeze through the tight pores of the network. This means that the gel ends up taking quite a while to respond to a change in conditions. Even more inconveniently, if we take a big lump of material and change conditions so that it starts to swell, then the outside ends up swollen before the solvent has a chance to get to the middle, and the resulting stresses often cause the material to crack.

But we know that as things get smaller, so the time taken for molecules to move by Brownian motion decreases as the square of the size. If it takes minutes or hours for solvent to reach the centre of a lump of gel the size of a tennis ball (say 6 cm in diameter), then it will take microseconds for a micron-sized piece of gel to swell, and nanoseconds for a single responsive molecule to respond.[18]

There have not been any demonstrations with synthetic materials of the use of shape transitions in gels that in any way approach the potential of this approach when systems are scaled down in size. But an observation made more than 300 years ago shows just how powerful these soft forces can be. In 1676, the pioneering Dutch microscopist Leeuwenhoek described in a letter to the Royal Society his observations on a tiny, single-celled animal connected by a stalk to a fragment of leaf: '. . .their whole body then lept towards the globul of the tayl . . . and un-wound again. This motion of extension and contraction continued a while . . .'. This contraction of the stalk (spasmoneme) of a vorticellid turns out to be, in proportion to its mass, the most powerful organ in biology, generating a power per unit mass ten times greater than a car engine. How does it work? The full details are not yet known, but the operating principle seems to be a gel consisting of negatively-charged polymer strands. When the stalk needs to contract, the gel is flooded with calcium ions, which neutralise the charge and bring about the collapse of the gel.

Conformational changes in proteins

Environmentally-driven shape changes in polymer molecules offer a way to couple chemistry to shape. Polymers show dramatic effects, but these effects are a bit unspecific. The master molecules for coupling a change in environment to a change of shape are proteins; we will see that this property offers the basis for how life tackles almost all of its most sophisticated tasks, from computing to movement.

Remember that a protein is composed of up to twenty different amino acids, linked together in a definite linear sequence. All of these amino acids interact with each other, and with the water around them, in different ways.

---

[18] In fact, these turn out to be underestimates, because when the gel is very small the response time is limited, not by how fast the solvent can get in and out, but by how long the polymer molecules take to change their shape.

Some amino acids are happy to be in contact with water, some are not. Some pairs of amino acids will want to stick to each other, others will be repelled by each other. When a protein folds to the well-defined, three-dimensional shape known as the native state, it is because, under the influence of the continuous shaking that the collisions by the surrounding water molecules inflict, the molecule arranges itself so that the water-loving amino acids are on the outside of the molecule, in contact with water, and the water-hating molecules are in the interior of the molecule, shielded from any contact with its wet surroundings. The mutually sticky pairs of amino acids stick together in a way that minimises the total energy of the molecule.

It is the mutual interactions—the patterns of mutual stickiness and repulsiveness between all of the possible pairings of the amino acids—that define the three-dimensional shape of the protein molecule. But the resulting molecule itself has a surface which has both cavities of different shapes and patterns of patches of different degrees of stickiness, and this combination of geometry and stickiness can lead to the formation of nooks that the right kind of molecule can fit into rather neatly. Imagine that our protein is sitting in a soup containing many different kinds of molecules, only one of which was the right shape and had the right pattern of surface stickiness to fit into the nook on the surface of the protein. As Brownian motion shakes up the molecules, all sorts of different kinds of molecules move in and out of the protein's sticky hole. But only the molecule that fits will stick; and, having stuck, it will resist being shaken off in response to the collisions of all of the other molecules, at least for a while.

This situation is often called 'molecular recognition'—a certain binding site on the surface of the protein picks out one molecule from the crowd around it, and holds it in a tight embrace. As we will see, it is this ability that underlies the ability of proteins to do all of their sophisticated tasks, from catalysing chemical reactions, to acting as molecular motors and molecular computers.

People have sometimes referred to a 'lock-and-key' mechanism as being the essence of the way a protein can pick out a molecule from a crowd. Just one particular molecular shape—the key—fits the shape of the cavity—the lock. While this analogy captures the importance of the match between the shape of the recognised molecule and the shape of the cavity, it does not capture the random nature of the recognition process. It is as if a burglar with a set of skeleton keys, rather than trying them out one by one to see which one fits his victim's door, could simply throw the whole bunch at the lock in the knowledge that the right key would magically insert itself.

The 'lock-and-key' picture also does not take into account the mutability of molecules like proteins, i.e. their essentially soft character. The matching is more like fitting a hand into a glove. A glove needs to be a good fit to the hand it fits on, but a really comfortable glove will stretch and mould itself to fit the hand that goes in it. In order to get into a harder glove—a protective gauntlet, for example—the hand itself will adapt in shape a bit. This is much more like the situation between a protein and the molecule it recognises; both molecules will slightly adapt their shape to get the best fit.

This simple idea—that a protein changes its own shape when another molecule binds to it—underlies much of the operation of life's soft machines. The following are three ways in which this basic idea can be used to make sophisticated machines.

- We can use the shape changes as the basis of a motor, converting chemical to mechanical energy.
- We can use shape changes in molecules that line the pores of membranes to give us valves, which open and close in response to this chemical signal.
- A motor and a valve can be combined to make a pump.
- In a molecule that has two binding sites, the shape change that occurs when a molecule docks in one binding site can mean that the other binding site becomes either more or less effective at docking its own target molecule. This is the basis for a kind of molecular communication—the protein is a detector, which, when it senses the presence of one kind of molecule, can broadcast that information by releasing another molecule. Cascades of signals and responses like this are known as 'signalling networks', and they are nothing less than sophisticated molecular computers, which process information about the environment and direct the organism to respond in some appropriate way.

We will talk more about molecular computing in Chapter 8. First, let us see how nature uses shape changes in proteins to make motors, pumps, and valves.

Power sources and energy budgets

If we do not eat, we die; this is a graphic illustration of the fact that life needs a constant supply of energy. It is not just a case of needing energy to do heavy physical work, like digging a ditch or carrying loads up a hill. Even if you have a completely sedentary existence then you still need to eat; an average man might need 5000 calories if he is doing a day of heavy work, but even if he stayed in bed all day then he would still need to use 1700 calories just to keep ticking over.[19] A surprisingly large proportion of this base energy requirement is needed to keep the brain and nervous system working, as we will discuss in Chapter 8.

All life needs energy, and the flows of energy around and between different organisms constitutes what is sometimes called the food chain. We meet

---

[19] The energy content of food is usually described in calories, a unit horribly familiar to many from diet books. Confusingly, a food calorie is the same unit as the chemists' kilocalorie, which is equivalent to 4184 Joules. So our bed-ridden man consumes about seven million Joules per day, which corresponds to an average power of 80 W—much the same rate of using energy as an ordinary light-bulb.

our energy needs by eating other organisms; if we are meat eaters, then these organisms themselves eat other organisms, but ultimately, at the bottom of the food chain, supporting everything else, are those organisms that take in energy from some primary source. This mostly means plants, which can convert sunlight into relatively-high-energy compounds like sugars. Although the vast bulk of life on Earth ultimately depends on the energy of the sun, it is not necessarily the case that this is the only primary energy source. There are whole ecosystems living in deep-sea vents whose ultimate source of energy is not the sun at all, but deep hydrothermal energy.

We will see later how plants use sophisticated molecular nanotechnology to harness the primary energy of the sun. Now I want to consider, not the ultimate source of that energy, but the currencies by which it is stored and circulated.

When we buy an energy drink or a bar of chocolate, to give ourselves a boost when we are feeling tired, we are relying on the fact that our bodies can readily extract energy from sugar. Table sugar—sucrose—is a small molecule, made up of carbon, hydrogen, and oxygen. By quite a complicated sequence of reactions our bodies can transform sugar into carbon dioxide and water, releasing a substantial amount of energy in the process. We can think of sugar as a store of energy. Macroscopically, we could release that energy by burning the sugar; sugar is a fuel. Of course, sugar does not burn that well—this suits its biological purpose, because we would not want to be in danger of bursting into flames. Sugars, and the molecules based on them, like starch and glycogen, are designed for long-term storage of energy. The biochemistry of the reactions which release the energy from sugar take too long for it be suitable as an instant source of energy.

We know about molecules that store a large amount of energy that is available for instant release—they are called explosives! The sort of molecules that are used as explosives and propellants—nitro-glycerine, for example—would not be suitable for use as energy stores in biological systems, for obvious reasons. The ability of such materials to release energy quickly, by a very simple chemical reaction, is what we need, but in these molecules the amount of energy that is released by each molecule—we can think of this as the packet of energy each molecule can deliver—is too large to be useful. The soft machines of biology cannot exploit a packet of energy that is hugely bigger than the energy of Brownian motion. Moreover, because this energy is so large the reaction is very difficult to reverse. It is easy to detonate explosives, but much less easy to make them.

The solution that biology has found is a molecule that shares with explosives a very straightforward chemical decomposition route, but in which the packet of energy released by the decomposition route is much more manageably small. This molecule is called adenosine triphosphate (ATP). It consists of the nucleotide adenosine—the same compound of an organic base and a sugar which forms one of the four components of DNA and RNA—loosely linked to three phosphate groups. The last phosphate group is particularly weakly attached to the molecule, and in the presence of water it will readily detach itself from the ATP molecule, releasing a packet of energy and leaving behind a molecule of adenosine diphosphate (ADP) and a free phosphate ion. The reaction can be reversed—with the addition of energy, and the removal of a water molecule, a free phosphate ion can be reattached to the molecule of ADP to give us ATP again. ATP is a reusable energy-storage molecule, keeping the energy available for instant use. If sugars correspond to the checking account of energy housekeeping, then ATP is the cash; convenient for instant use, but not suitable for long-term storage.

We are talking here about molecules storing energy, but how do they actually do this? There are two ingredients to this, one of which is straightforward and one is not. First of all, there is an energy that arises from the fact that there is a force between an ADP molecule and the phosphate ion as the ion is separated from the ATP. For the reaction of ATP, which results in the production of a free phosphate ion and the absorption of a molecule of water, energy first needs to be put into the system to start the reaction off, but after that the phosphate ion is physically repelled from the remaining ADP molecule. This repulsion can be used to do work, and the amount of work thus done outweighs the energy that was needed to be put in to start the reaction off (which chemists call the activation energy). So we have a certain quantity of what is essentially mechanical energy released, which we could in principle calculate if we knew all of the forces between the parts of the ATP molecule as it split up.

But in addition to this there is another factor—entropy. We saw in the last chapter that there is a natural tendency for things to get more disordered, which is formally expressed by the second law of thermodynamics. The entropy of a single ATP molecule in a certain volume is lower than the entropy of one ADP molecule and one phosphate ion in the same volume; if the ADP and phosphate are constrained to be in the same place then the system is necessarily disordered.

The importance of entropy is that one can store and convert energy without any chemical reactions, simply by sorting molecules. If we have two compartments, one containing pure water, the other containing a concentrated solution of some chemical (salt, perhaps, or sugar, any solute will work), then this is a situation of lower entropy than if we allow the two compartments to mix, so we end up with both compartments containing the same lower concentration of the chemical. Before mixing, the system has a higher free energy than after mixing, and we can use that change in energy to do useful work. The most usual way of thinking about this is that, if the two compartments are separated by a membrane which allows water to pass, but prevents the solute from going through, then the more concentrated solution develops a pressure—an osmotic pressure. If the volume of the compartment holding the concentrated solution is allowed to increase, then this pressure will do work.

Close to my house in Derbyshire there are many low cliff-faces, the remains of quarries from which the rough sandstone was taken between the Middle Ages and the early nineteenth century to make millstones for grinding flour and grindstones for sharpening the knives made in nearby Sheffield. Before the introduction of powered stone-cutting saws, the rock had to be broken away from the quarry face and shaped by hand. You can still see the evidence of one of the tricks the old quarrymen used to do this; along the top edges of some of the big, car-sized blocks they left behind are rows of square, in-cut grooves. What they would do to split a block from the quarry face is chisel a series of rectangular holes in a line parallel to the edge of the rock face. Into each of the holes they would hammer a wooden wedge. Water would be poured on the wedges, which were kept wet. As the wood swelled in the water, the force would break the block off from the quarry face. It seems unlikely that a combination of water and wood can break hard stone, but that is the quiet power of osmotic pressure.

In biology, this osmotic effect is used to store energy by creating high concentrations of hydrogen ions inside a membrane. This osmotic energy store is a key intermediate in the energy economies of all living things; energy gained by breaking down food molecules or from sunlight is used to pump hydrogen ions across a membrane, and the osmotic pressure of the concentrated hydrogen ions in these compartments is used to drive the machines that convert ADP to ATP, providing the energy source for all of the cell's other operations.

Protein-based linear motors

The last industrial revolution produced more than its fair share of stories of pathos, proud men and women whose livelihoods were taken away by mechanisation, and the social dislocation it caused. These stories are mostly told in terms of heroic futility. The Luddite movement has survived only as a term of abuse for people who do not like progress. One story still retains its heroism, however, the tale of John Henry, whose personal battle against mechanisation is told in the folk song 'Gonna die with my hammer in the hand'.[20] John Henry worked building railroads in the USA in the 1870s and had a drilling competition with a steam hammer, which he won, but at the cost of his own life.

Muscles are a remarkably effective source of power; for a machine that is made of meat, it does surprisingly well. The specific power of striated muscle is two hundred Watts per kilogram, compared to only three hundred for a petrol engine. Given that a nineteenth-century steam engine is certain to have been much less efficient than a modern petrol engine, it is ironic that in terms of thermodynamic efficiency and power/weight ratio, John Henry was almost certainly more efficient than the steam engine that killed him.

Suppose that we could shrink John Henry and the steam hammer down to nanoscale size—how would the competition end then? Almost certainly with a victory to John Henry, because steam engines, like all heat engines, work less and less well as the size scale is shrunk. But muscles would work as well on

---

[20] As recorded by many artists under a variety of names. The definitive version is recorded under this title by The Williamson Brothers and Curry (Anthology of American Folk Music, Smithsonian Folkways Recordings), but other versions have been done by artists from Johnny Cash to Memphis Slim.

the nanoscale as at our human scales. In fact, what we have in a muscle is many nanoscale motors, each the size of a single protein molecule, all working together to produce their macroscopic effect.

The working core of a muscle consists of interlocking arrays of filaments. There are two types of these filaments. The thin filaments are composed of chains of molecules of a protein called actin. Between the thin filaments are the thick filaments, which are bundles of another protein called myosin. Each of these myosin molecules has an arm which protrudes from the central bundle, and on the end of the arm are a pair of what we might as well call hands. When the muscle is activated, what happens is that the hands on the ends of the myosin arms grab the nearby actin filament and give it a pull, before letting go. The combination of many pulls from many myosin hands causes the thick filament and the thin filament to move with respect to each other, and the result of the many thick and thin filaments in the muscle moving over each other like this is to make the muscle contract.

Underlying the mechanism of this remarkable process are those central features of the nanoworld, namely Brownian motion, conformational change, and stickiness. Both ATP and ADP bind to the same place in the head-group of myosin, but, because these two molecules have slightly different shape, the shape of the myosin molecule itself is distorted in different ways when each molecule is associated with its binding pocket. A consequence of this change in shape is that the degree to which another part of the myosin molecule sticks to the actin filaments changes. The hydrolysis of ATP to ADP thus leads to the changes in the shape and stickiness of the myosin which underlie the action of the motor. Let us consider this in a little more detail.

We can identify four distinct states of the myosin molecule in its association with ATP (see figure 6.3). Having bound the ATP, the myosin catalyses ATP to give ADP and a phosphate group. With both an ADP molecule and the phosphate group in myosin's binding pocket, the myosin molecule is extended in shape and very sticky. When Brownian motion brings the myosin in contact with the actin filament, it sticks to it. This leads to a slight change in the shape of the binding pocket, which makes it less congenial for the phosphate group, which leaves. With only ADP present in the binding pocket, the myosin now undergoes quite a big change in shape. A small change of shape at the binding site is translated by lever action to rather a large movement of the part of the molecule stuck to the actin; this is the power stroke, in which the myosin gives the actin filament a yank resulting in about 10 nm of motion.

It is worth pausing to remind ourselves of exactly what is going on in this power stroke. What actually causes the myosin molecule to change shape? It is the collisions of the surrounding water molecules that actually exert the force, with the changed pattern of internal stickiness that arises when the phosphate ion leaves then biasing the random internal motions of the myosin, so that it moves towards its new equilibrium shape.

How is the cycle completed? The myosin molecule, in its sticky and contracted state, has an ADP molecule in its binding pocket. The binding is relatively

(a)   (b)

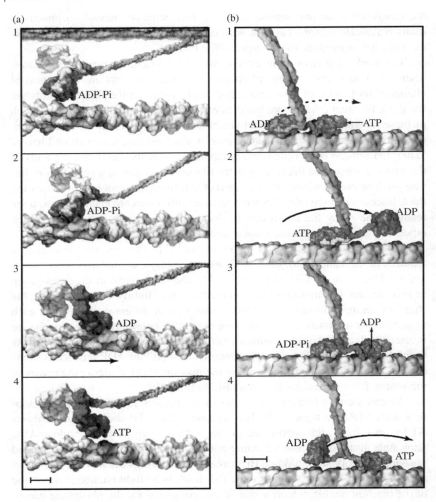

Fig. 6.3 The motor cycles of (a) myosin, and (b) kinesin showing the cyclic binding of adenosine triphosphate (ATP), its hydrolysis to adenosine diphosphate (ADP) and a phosphate ion (Pi), and the release of the ADP molecule. The scale bars are 6 nm (A) and 4 nm (B). Reprinted with permission from R. D. Vale and R. A. Milligan. *Science* **288** (2000) 88. Copyright 2000 AAAS.

weak, and when Brownian motion shakes off the ADP, it will tend to be replaced by a more-strongly-binding ATP molecule. In response to binding this differently-shaped molecule, the myosin molecule once again changes shape, back to its original extended state, and detaches from the actin filament, taking us back to the beginning of the cycle.

There are many other types of linear motors in biology, whose cycles differ in detail from the muscle motor just described. But the physical principles are the same—conformational change and switchable stickiness coupled to the catalysed hydrolysis of ATP, with the shape changes being driven by Brownian motion. They share the characteristics of a Brownian motor; most importantly, their operation is statistical rather than exact. It is all too easy, when one looks at illustrations in papers or textbooks of the stages in these motor cycles, to visualise the sort of cycle we are familiar with from steam engines or petrol engines, in which each stroke has precisely the same length and each cycle is identical to the one that went before. But biological motors are much more haphazard affairs; detailed studies at the single-molecule level show that the strokes in successive motor cycles are, in many cases, far from constant in length; in fact, they are not always even in the same direction. One step back and a few steps forward is how a biological motor works; and the steps forward often are of different lengths. But while the individual steps are faltering, a muscle has so many of these acting in concert that the result can be very powerful.

Rotary motors in nature

Wheels are very rare in nature at the macro-scale, and the reason for this has attracted a certain amount of learned comment. It should not really be a surprise, though; British people of my age will remember a television science fiction series called 'Doctor Who', featuring a race of terrifying evil aliens called the Daleks. These resembled giant dustbins on castors; they rolled around pointing their built-in death-ray guns at people, repeating the word 'exterminate' in robot tones. I was terrified by them as a child, but their spell for me was terminally broken by a memorable cartoon showing a group of Daleks at the foot of a short flight of steps. 'Well, this certainly scuppers our plan to take over the Universe', was the caption.

Rather surprisingly, rotary motion is important in biology at the nanoscale. Those bacteria that are able to swim about do so using a rotary motor which uses the stored energy of a hydrogen ion gradient to rotate a long, thread-like flagellum. The resulting corkscrew-like motion is quite an efficient mode of propulsion in the viscosity-dominated nanoscale environment.

The rotary motor is mounted in the bacterium's cell membrane; the rotating parts consist of a series of circular rings of protein, which are driven to rotate by an outer fixed ring of another type of protein molecule. The driving force for rotation is provided by the osmotic pressure arising from there being a higher concentration of hydrogen ions inside the bacterium than outside.

Hydrogen ion gradients also drive another remarkable rotary engine that all life-forms rely on. This is the assembly of proteins called ATP-synthase that adds a phosphate ion to ADP to yield the high-energy molecule ATP. Once again, this is a machine that is mounted in a membrane. Inside the cells of animals like ourselves, for example, are organelles known as mitochondria, consisting of a highly-convoluted membrane separating an interior compartment from the rest of the cell. As we metabolise food molecules like glucose, the energy produced is used to pump hydrogen ions across the membrane into the mitochondrion. The flow of these hydrogen ions back across the membrane through ATP-synthase causes part of the synthase assembly to rotate, and this rotational energy is used to add phosphate ions to ADP to make the ATP.

The way this process of energy conversion works is still not yet understood in the most complete detail, but the structure of this remarkable machine has now been determined. There are essentially three parts: a ring of protein molecules which spans the membrane, a stalk which is mounted like an axle in the transmembrane ring, and a bulbous head. The head is joined to the transmembrane ring by a fixed arm, and the other end of the stalk is inserted into the head's centre. As hydrogen ions diffuse through the assembly of the transmembrane ring and the central axle, conformational changes in the ring and axle proteins cause the axle to rotate. The rotation of the stalk inside the head likewise leads to conformational changes in ADP binding sites, leading to the synthesis of ATP.

ATP-synthase is one of the key pieces of nano-machinery that makes life possible. Essentially, the same machine occurs in animals, in plants, and in the simplest bacteria. Some time in the early history of life, evolution perfected this exquisite mechanism, exploiting conformational change and Brownian motion to interchange different types of chemical energy through the medium of mechanical energy. Despite all of the developments in complexity that separate a complicated, multi-cellular animal and a simple bacteria, this vital component remains essentially unchanged.

Synthetic nanoscale motors

Evolution had a few billion years to develop the principle of molecular motors—the coupling of conformational change and chemical reaction that converts chemical energy into mechanical energy in conditions of constant temperature. It is perhaps not surprising that our own efforts have been much less successful so far. We are still quite a long way from making a synthetic, reciprocating motor than can take a source of chemical energy and do something useful with it (like propel a cargo, for example) on the nanoscale.

What we can do quite successfully is take apart the components of living cells, isolate natural motors from them, and reassemble them in a synthetic context. This was first done from motives of pure science—to allow controlled experiments with which to better understand their modes of operation. For example, direct proof that the operation of ATP-synthase involves rotary motion was obtained by Hiroyuki Noji and Masasuke Yoshida. They were able to isolate these molecular machines from bacteria, anchor them to a surface, and then attach protein threads, 4 microns in length, to their rotors. When they added ATP fuel to the solution, they could watch the threads rotating in a light microscope.

But having learnt how to handle the components of biological motors, it is a small step to proposing to use them in entirely artificial situations.

Collaborating with the physicist Harold Craighead at Cornell University, Carlo Montemagno combined biological and inorganic nanostructures to create a hybrid device. To start with, they used electron beam lithography to make a surface coated with a series of pillars, each about 100 nm in diameter and 200 nm high. They also used electron beam lithography to make tiny metal propellers, each of which was 150 nm in diameter and about a micron long. The idea was to attach a single ATP-synthase molecular motor by its static part to each pillar, and then to mount the propellers on the ATP-synthase rotors.

It is conceivable that one could individually mount each rotary motor on a post, using, for example, atomic force microscopy. But Montemagno and Craighead chose the much more powerful approach of self-assembly. By genetic engineering, it was possible for Montemagno's group to create ATP-synthase with different sticky attachment groups on the static part and on the rotor. The sticky groups on the static part of the motor were designed to stick to the surfaces of the posts. Another sticky group was attached to the rotor; the metal propellers were themselves coated with a protein designed to bind to this. Now, assembly of the hybrid machine was simply a question of putting the components together and waiting for them to find each other by Brownian motion and stick together in the right order. The test that these organic/inorganic hybrids had correctly put themselves together was to add ATP fuel; the result was that the propellers duly started rotating.

But taking apart nature's finely-optimised mechanisms and reassembling them is not quite as satisfying, or as instructive, as making something entirely from first principles. Some attempts to achieve the design and synthesis of a molecular motor from scratch are now going on.

Inspired by Nadrian Seeman's use of the highly-specific self-assembling properties of DNA to make complex structures, Andrew Turberfield, at Oxford, in collaboration with Bernie Yurke at Lucent's Bell Laboratories, has used the base-pairing mechanism to convert chemical energy into the motion of constructions made of DNA. The motor uses four designed sequences of DNA, namely Q, L, L', and M. L and L' are two complementary sequences that together would form a double helix, but we start by putting together Q and L', which form a shorter double helix with a protruding loop. We can think of the motor cycle as the reaction QL' + L → LL' + Q; because more base-pairings are made in the final double helix LL' than in the initial helix and loop QL', this releases stored energy. What is needed to make the reaction proceed is the fourth DNA strand M, which catalyses the reaction. In the process of carrying out this catalysis, it cyclically changes its shape. If we could couple this shape change to some mechanical component then this would be able to do useful work.

These DNA motors use materials from nature, but in quite a different context. They are true isothermal motors, using Brownian-motion-driven conformational change as the power stroke, and they capture extremely well the principle of biological motors in which a molecular shape change is coupled with the catalysis of a chemical reaction. They are a long way from practical application, however. The materials are still rather expensive, and the fuel is even more exorbitant. (Here 70 kcal/mol is the energy release per cycle, which needs an input of two 70-base DNA molecules.) The advantage of using DNA, though, is that these molecules are currently the only ones whose self-assembly we understand in anything like enough detail to be able to design catalysts and predict shape changes with any degree of certainty.

My own research group at Sheffield, in collaboration with my chemist colleague Tony Ryan, are taking the low road towards a molecular motor, using very cheap materials. We were inspired by the work of Ryo Yoshida in Japan, who coupled the type of responsive gel that we introduced earlier in the chapter with a non-linear chemical reaction, in which the chemical environment spontaneously oscillates as the ingredients of the reaction are consumed. His group uses the most famous of all oscillating reactions, the Belousov–Zhabotinsky reaction.

Oscillating chemical reactions are fascinating but notoriously difficult to understand, because the real systems that have this property tend to have many components and involve many different chemical reactions. We can get a rough idea of what is going on by considering a highly-simplified scheme, which goes something like this. Suppose we have two chemicals, A and B, which react together to make a product C. In the absence of B, another slow reaction converts C back to B again, but if plenty of B is already present then C is converted to D, which is the final product of the whole reaction. The intermediate product C has an important property: it catalyses the reaction by which it is itself produced. We are going to carry out the reaction by continuously feeding the material A (which we can think of as a fuel) into a pot which already contains B; as the reaction proceeds we are going to remove the final product D, which we can think of as the exhaust. When we begin, by turning on the flow of A into our pot of B, the reaction starts off slowly, because no catalyst is present. But as the reaction proceeds it produces its own catalyst, C, and so the reaction races ahead. The more catalyst C is produced, the faster the reaction goes, producing yet more C, which makes the reaction go even faster, and so on. There must be a limit to this acceleration of the reaction; eventually we must use up all of the material B. At this point the reaction slows right down, and the second, much slower reaction takes over. As this happens, some of the catalyst C is converted back to B again. Eventually, enough B is being produced for the reaction of A and B to form C to start over again, and the cycle starts once more.

Looked at from the outside, this is a fairly conventional reaction, with a higher-energy fuel A being converted into a low-energy product D. But the combination of autocatalysis and feedback means that the concentrations of B and C oscillate with time. In the classical Belousov–Zhabotinsky reaction (which, it should again be stressed, is very much more complicated than this simple cartoon of an oscillating reaction), these oscillations are made visible by a colour change. It is a good lecture demonstration; without any intervention one sees a beaker of chemicals change colour alternately between red to blue every few minutes.

We adapted Yoshida's chemistry by using a different oscillating chemical reaction and a different polymer. In our system, what oscillates is the acidity of the solution. Our solution goes from being roughly neutral to quite acidic—below pH 4—every few minutes. We immerse in this reaction bath a polymer gel similar to the type used in disposable nappies—polymethacrylic acid. As we discussed above, this is a weak acid; when immersed in a neutral solution it ionises, is easily soluble in water, and it swells. In acid it remains uncharged, and, being rather hydrophobic in this state, it collapses. When we combine the oscillating chemical reaction with this responsive polymer we end up with a piece of jelly that spontaneously throbs as the acidity of the bath changes. As the gel expands we can make it exert a significant force. What we have here is an isothermal engine; it is directly converting part of the energy that is released when the fuel molecule is converted to the exhaust product into mechanical work.

What is attractive about the throbbing gel, from the point of view of making a molecular motor, is that the macroscopic motion of a centimetre-sized lump of gel reflects the conformational changes of each and every polymer strand inside it. This means that what works at the macroscopic level should work for a single molecule. So how in practise can we scale down the throbbing gel to give us a throbbing molecule, preferably one we can hold by both ends and feel the force?

One way we can do it is by creating a surface coated by a single layer of polymer molecules, with each molecule attached by one end, like the bristles of a brush. If we can then grasp the other end of the molecule, for example with the tip of an atomic force microscope, then we could feel the pulling and pushing forces that the molecule exerted as the pH cyclically changed. The conceptual problem with this scheme to make a chemical motor is that there is no direct coupling between the conformational change and the chemical reaction, which makes it very inefficient. Most of the reaction takes place far away from the molecule and its energy is wasted.

Andrew Turberfield's motor is elegant, but extremely expensive in materials and fuel. Ours is cheap, but wasteful, because the catalysis of the reaction is not directly coupled to the shape change of the molecule. What we are all groping after is a workable and efficient scheme for coupling catalysis of a chemical reaction with conformational change in a way that is inspired by the operation of biological molecular motors, but which uses entirely synthetic components. If we can do this, then we will have found the first example of a way to power a soft machine.

I am certain that a research group somewhere in the world will soon work out the best combination of the high-road and low-road approaches to a synthetic molecular motor, and make one that can do something useful (and naturally, I hope that group will be mine).

Mechanisms and machines

Gears and levers

When the industrial age began three hundred years ago, many skilled crafts-men rapidly lost their livelihood, as factory-based mass production rendered

Fig. 6.4 Medieval macrotechnology. A water pump driven by a human-powered treadmill. G. Agricola. *De re metallica*. Translation by H. C. Hoover and L. H. Hoover (1912), reprinted by Dover, New York (1950).

their skills obsolete. But some craftsmen found themselves in great demand. Millwrights, who understood how to convert the power generated in a windmill or a watermill to rotate the heavy millstones used to grind corn, found themselves a lucrative new occupation designing and building the systems of pulleys, gears, and cogs that were needed to translate the raw power generated by the first giant steam engines into a form in which it could be used by the new machines. Our mental picture of machinery is as much dominated by images of the assortments of cogs and ratchets, and gears and flywheels that are needed to mediate the output of prime movers like steam engines as by the engines themselves (see figure 6.4). We imagine these mechanisms as having almost sinister power: 'wheels within wheels', we sometimes say, in referring to complex machinations and sinister conspiracies.

Before the digital age, some of the most impressively intricate technology that an everyday person might encounter was the mechanism of a watch. Watchmakers have learnt to create motions that combine great intricacy and accuracy with very small size. Smaller sizes still are possible using the planar fabrication technologies that we discussed in Chapter 3, in the context of micro-electro-mechanical systems (MEMS). In demonstrating the potential of MEMS as a technology, some of the most spectacular creations have been sets of gears whose total dimensions have been measured in hundreds or even tens of microns (see figure 6.5).

It is very natural to ask how small this process of scaling down can go. Is it possible to design and build a working assembly of gears and cogs—a gear box, say—whose overall dimensions are smaller than one micron? For some

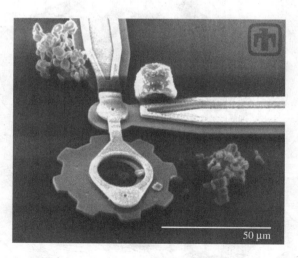

Fig. 6.5 Microtechnology made by planar silicon processing techniques. The drive gear chain and linkages for a micro-machine, with a grain of pollen (top right) and coagulated red blood cells (lower right and top left) to demonstrate scale.
Courtesy of Sandia National Laboratories, SUMMiT™ Technologies. www.mems.sandia.gov

writers, overcoming this problem is key to realising the potential and the promise of nanotechnology. Drexler, and writers who have followed him, have thought hard how we might engineer cogs and gears at the atomic level. This kind of engineering, so far, has been virtual rather than real, with the results appearing on a computer screen rather than as physical artefacts. We can use molecular modelling programs—the computer versions of the models that students used to construct molecules from rods and coloured balls. With these modelling programs, we can easily design structures that at least obey the basic laws of chemistry. With slightly more sophisticated computer programs, using as input the known forces between atoms in molecules, we can begin to test the mechanical properties of our putative components. In this way, ingenious designs for tiny cogs and bearings, and all sorts of other mechanical components, have been proposed and studied (see figure 6.6).

Most of this virtual engineering has been done using structures made out of carbon atoms (see figure 6.7). It is appealing to some to think that this most common and widely available material might be the basis of such a spectacularly transforming technology. Nanotechnology is often written about as a kind of new alchemy, as a way of transmuting common things into whatever is most precious, so it is perhaps not surprising that early nanotechnologists alluded to a real piece of alchemy. As everybody who remembers their school chemistry

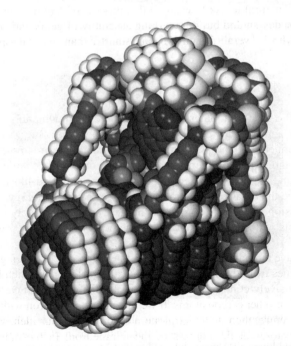

Fig. 6.6 A design for a fine-motion controller for molecular assembly by K. Eric Drexler. © Institute for Molecular Manufacturing. www.imm.org

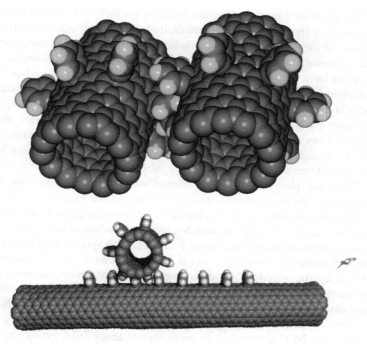

Fig. 6.7 Computer simulations of gears and shafts made from functionalised carbon nanotubes. From A. Globus, C. W. Bauschlichers, Jr, J. Han, R. LJaffe, C. Levit, and D. Srivastava. *Nanotechnology* **9** (1998) 192.

knows, the common element carbon exists in two forms. By far the most common is the grey, soft material called graphite, from which the lead in pencils is made, and which in impure form constitutes soot. But at conditions of very high temperature and pressure, this unpromising raw material transforms to the rare and expensive gemstone diamond, one of the hardest materials known to man. Drexler suggested that carbon in its diamond-like crystal structure would be the ideal construction material for nanotechnology, an idea that finds expression in the title of one of the best and most widely-read science fiction novels with a nanotechnology theme, *The diamond age* by Neal Stevenson.

As we have already seen in Chapter 4, the difference between graphite and diamond comes from the way in which the carbon atoms of which both materials are exclusively composed are arranged. In diamond, each carbon atom is attached to four other carbon atoms. If we imagine a tetrahedron with one carbon atom at the centre, then that central atom is joined by four identical bonds to four carbon atoms at the corners of the tetrahedron. In the diamond in an engagement ring, vast numbers of carbon atoms are perfectly arranged in a network in which every carbon atom is linked to four neighbours in this way.

What virtual nanotechnologists, working on their computer screens, have imagined are finite clusters of tetrahedrally-bonded diamond. Of course, at the surfaces of the clusters the carbon atoms do not have their full complement of four neighbouring carbon atoms, but the unsatisfied surface bonds can be occupied by a hydrogen atom. These structures, clusters of carbon bonded with diamond-like bonds but with their rough edges smoothed off, as it were, with a layer of hydrogen, are known as diamondoids. On the computer we can construct diamondoid components of any shape we want, gears, cogs, or shapes of any complexity we desire. The only problem is that, while we can draw these structures on the computer screen easily enough, no one has any idea how to make them in reality.

There are plenty of times in science when fortune favours the wildest speculators, and nano-enthusiasts were as stunned and delighted as the rest of the scientific community by the chance discovery of fullerenes and nanotubes; these are a series of stable, nanoscale clusters of carbon atoms, with beautiful, almost architectural, designs. As we discussed in Chapter 4, Harry Kroto, from Sussex University in the UK, and Richard Smalley at Rice University, discovered a structure of carbon consisting of a closed sphere made up of five- and six-membered rings, which they named 'buckminsterfullerene' after the innovative architect and designer Buckminster Fuller, who proposed similar structures as efficient ways of constructing ultra-lightweight domes. Sumio Iijima, of the NEC laboratories in Japan, subsequently discovered carbon nanotubes, which consist of sheets of carbon rolled into cylinders, with closed ends. The most perfect form of nanotubes, in which a single sheet of carbon is rolled round to form a cylinder and each end is capped with half a buckminsterfullerene molecule, was identified in 1993.

These beautiful nanostructures seem to indicate that the diamondoid structures of the extreme nanotechnologists might not be so fantastic after all—here was the proof that beautiful constructions could be put together out of carbon atoms. There was just one snag—these structures had nothing to do with diamond. Remember that, in diamond, each carbon atom is joined to four others. But in fullerenes and nanotubes, each carbon is only bonded to three other carbon atoms. To get a nanotube, imagine a sheet of chicken wire. It is made up of hexagons; from each junction point, three pieces of wire go out in a plane. If we cut a strip out of the chicken wire, roll it into a cylinder, and join the wires, then we have the tube. To make the caps for the ends of the tube, we need to replace some of the hexagons by pentagons, so that the sheet no longer lies flat, but curves around in two directions to make hemispheres.

The bonding of the carbon in fullerenes and nanotubes is not at all like the bonding in diamond; actually, it is the same as the bonding in graphite. In this soft, grey material, each carbon atom is bonded to three others, and the resulting sheets of hexagons are piled flat on top of each other, like a gigantic pile of old newspapers. Does this mean that fullerenes and nanotubes are going to be useless for making the components of our nano-machines—rather than the adamantine strength of the diamondoids, do we not end up with the sooty softness of graphite?

The surprising answer is no. The strength of the bond between neighbouring carbon atoms in diamond is very strong, and it is this that accounts for the very high strength and stiffness of diamond compared to virtually all other materials. But the strength of the carbon–carbon bond in graphite is even stronger still; in fact, it is the strongest chemical bond we know about. The apparent softness of graphite is deceptive; if we were able to grip two ends of the same sheet of carbon atoms and pull, then this individual sheet would be very strong. But while each sheet of carbon atoms is strong, the force holding the sheets together is weak, and it is the ease with which the sheets slide past one another that leads to the softness of graphite. But since a nanotube is essentially a rolled-up single sheet of graphite, if we pull on a nanotube then we are guaranteed to be pulling directly on the strong carbon–carbon bonds.

In the world of fullerenes and nanotubes, then, we have a synthetic chemistry that allows us to make nanoscale clusters of carbon atoms which have well-defined architectures and which are very strong. Does this show us the way forward to make the computer simulators' dreams of molecular-scale cogs and gears into real mechanisms and machines? My belief is that there are three barriers to overcome if we want to get from our current, rather rudimentary competence in synthesising and manipulating fullerenes to making a mechanism of the type envisaged by the nanotechnology enthusiasts. Two of these barriers are challenging but not insuperable. But the final barrier is more serious, and I suspect it will scupper the entire enterprise.

The first barrier to overcome is in terms of finding synthetic routes to make structures more complicated than the spheres, ovoids, and tubes that constitute the current fullerene repertoire. Will it really be possible, for example, to take a nanotube, and attach a series of tooth-like molecules around the circumference of the molecule to make a nanoscale cog-wheel? Inorganic chemists are now learning how to attach chemical groups onto the surface of nanotubes, but the degree of control that is now possible leaves a lot to be desired. Nonetheless, in a field that has already yielded so many surprises and so much serendipity, it would be foolish to underestimate what chemists will make possible.

Assembling the components of one's machine is not going to be easy, either. It is true that one can move things around with the tip of a scanning tunnelling microscope or a scanning force microscope. The current state of the art is to prod a nanotube to make it lie across a pair of electrodes. Will it ever be possible to pick up a gear wheel, thread it onto a nanotube, and mount the resulting assembly onto a frame? With our current technology this kind of task would be like trying to reassemble a watch mechanism wearing winter mittens. Nonetheless, scanning probe microscopy is still relatively young, and we can expect refined instrument designs optimised for this kind of manipulation before too long.

What I think is going to thwart the vision are some much more fundamental problems, ones which it is simply not going to be possible to engineer around. These are related to the fundamental nature of the nanoworld, and the dominance of Brownian motion and surface forces. The continual shaking that every nanoscale structure is subject to due to collisions with the surrounding gas or liquid molecules, and the large forces that will stick any surfaces that come into contact irreversibly together are inescapable obstacles to the kind of precision nano-engineering that would be needed to miniaturise systems of gears and cogs to the nanoscale.

Too few people appreciate how important to the development of technology has been the ability to manufacture parts whose dimensions are controlled to very high degrees of tolerance. By 1792, James Watt was able to improve the efficiency of steam engines from the value of 1.4%, achieved by Smeaton in 1774, to a value of 4.5%. This improvement came from Watt's invention of a separate condenser, but it was only made possible by improvements in the precision with which metal could be worked. Watt's early designs were thwarted because cylinders made by the best available workmen varied in diameter by 3/8 of an inch. John Wilkinson, in Shropshire, devised a new boring machine that could make a 72 inch cylinder with an error less than 'the thickness of a thin sixpence', which made Watt's design viable. Expressing these tolerances as a percentage of the overall dimension, we see an improvement in achievable tolerances from about 1% to 0.05%.

In the middle of the nineteenth century, Joseph Whitworth, who is now most famous for having standardised screw threads, dominated mechanical engineering. He introduced many new and improved machine tools and measuring instruments, as well as an attitude that placed a high premium on precision. Using his equipment, it was possible for a craftsman to manufacture components to a tolerance of 1/10000th of an inch—in percentage terms, roughly 0.001%. It was this improvement in manufacturing accuracy that made mass production possible. In the past, in order to make a complicated piece of machinery like a gun, each part had to be individually made and fitted together by a skilled craftsman equipped with a file. Whitworth's increase in precision made it possible to make many parts whose dimensions were closely controlled enough to make them interchangeable, and it was only this development that made it possible to mass produce objects.

On the face of it, one might think that there can be no problem regarding tolerances in nanotechnology—after all, one of the benefits of engineering with atoms is that quantum mechanics fixes their bond lengths. If I have a molecule of buckminsterfullerene, then I know that it has exactly sixty carbon atoms. It is not a question that most molecules have sixty atoms, a few have fifty-nine, some others have sixty-one, and on Friday afternoons some get made with fifty-five and are thrown out by quality control. If it has not got sixty atoms, then it is not stable and it is not buckminsterfullerene. Since the size of each carbon–carbon bond has an exactly known length, we know exactly how big a fullerene molecule should be. If we are engineering with individual atoms, then there should be no place for error in the size of our structures.

But this reassuring view does not take into account of Brownian motion, that inescapable feature of the nanoworld that we introduced in Chapter 4. No structure is absolutely rigid, not even a highly-symmetric and strongly-bonded molecule like buckminsterfullerene. If our nanoscale mechanism is immersed in water at room temperature, then it is going to be bombarded by colliding water molecules, and it will bend and flex in response to these bombardments. We can estimate with some precision just how much play in the structures this will result in.

For simplicity, let us imagine using a carbon nanotube as an axle, threaded between two bearings some distance L apart. If we were to press on it at its midpoint, then it would deform slightly, just like a plank of wood suspended between two bricks. The deformation stores elastic energy. As a result of collisions by surrounding water molecules, the nanotube will be continuously flexing in such a way that the average energy stored is one-half of the product of the temperature (on the absolute scale) and Boltzmann's constant. It is the size of these flexings that define the effective tolerance, not the accuracy with which we can control the size of the components. Using a reasonable estimate for the rigidity of a carbon nanotube, one can show quite easily that, for a nanotube held between bearings 10 nm apart, the average flexing half-way between the bearings is about one-tenth of a nanometre, more than ten per cent of the radius of the tube.

Is there some way we can get round this Brownian motion problem? The more rigid the material, the less the play will be—but nanotubes are close to the most rigid materials we know about. Could we somehow avoid the problem by excluding any water or gas molecules, so there is nothing to collide with our components? As we discussed in Chapter 4, Brownian motion does not depend on there being physical particles to make the collisions—even if our mechanism is in a perfect vacuum, photons being emitted and absorbed from the surfaces will lead to a play in the mechanism of exactly the same size, as the size of these fluctuations depends only on the temperature that the mechanism is held at. So, only by the impractical expedient of cooling our system close to absolute zero would we recover the near-perfect engineering tolerances that we might think result from construction using individual atoms as units.

The strength of surface forces is going to provide problems that will be just as difficult to overcome. We saw in Chapter 4 that the forces between identical surfaces become very strongly attractive for dimensions measured in nanometres. This is already proving a problem for the development of MEMS systems, where a phenomenon known as 'stiction'—a combination of sticking together and friction—causes serious obstacles to keeping moving parts free on the micron scale. As we saw, the ultimate origin of these attractive forces is in the fundamental properties of the quantum fluctuations in the electromagnetic field in free space—the Casimir effect—and this is going to be very difficult to engineer around.

Pumps and valves

Biology is intrinsically wet; the operation of our bodies depends on the trouble-free pumping of fluids, and a surgeon probably has more in common with a plumber than a car mechanic. On the nanoscale of a bacterium, the bulk flow of fluids is less important, not least because of the dominance of viscosity on these length scales, but moving dissolved molecules in and out of compartments is the cell's core business. The key mechanisms for biology, then, are not so much gears and levers as pumps and valves.

Artificial nanotechnology, too, is going to find some important applications for tiny pumps and valves. One very fast growing area of technology is in the kind of micro-chemical systems known as the 'lab on a chip'. The idea here is to carry out simple or complex chemical reactions—diagnostic tests, for example—not at the laboratory scale, in glass flasks and test-tubes joined together by tubes and pipes, but in tiny systems of reactors and channels etched into the surface of a piece of silicon or glass by the sorts of planar fabrication techniques we talked about in Chapter 3. The ultimate goal of this kind of system could be to make a complete diagnostic and therapeutic centre in a package small enough to be implanted in a human.

It is easy to see how this might work to alleviate those diseases that occur when the levels of certain biochemicals come out of balance in the body and need to be corrected artificially. Many women, suffer from the condition hypothyroidism, in which the thyroid gland, for one of a number of possible reasons, fails to produce enough of the hormone thyroxine. One hundred years ago, this disease was incurable and ultimately fatal, but now it is controlled by taking a supplement of artificially produced thyroxine. Now, the way the condition is managed is that every couple of months the patient makes a trip to the doctor's office, where a blood sample is taken and sent off to a laboratory in a nearby city for testing. A week or two later, a postcard with the result on is sent back to the doctor, who then decides whether the dose of thyroxine needs to be increased or decreased. How much simpler and more accurate it would be if the processes of testing the level, calculating the proper dose, and releasing it into the bloodstream were all carried out automatically.

There are some difficulties with microfluidics that arise from the basic changes in the way matter behaves on small scales. The key issue, as I mentioned above, is the increased importance of viscosity at small scales. This has one obvious effect: it is very much more difficult to push fluids around small pipes than it is through big ones. There are some less obvious consequences, too. It is very much more difficult to get fluids to mix on a small scale than on a large one. When we put milk in our tea and give it a quick stir, we are relying on the fact that the spoon produces a complex pattern of flow, whose characteristic is that there is rotational flow on lots of different length scales. On top of the general circular flow around the mug, lots of smaller whirlpools are generated at the edge of the spoon, and if we were able to look closer then we would see tiny whorls being spawned from the whirlpools. The effect of this is that the splash of milk we have added is strung out and stretched and folded back until in a few moments it is completely mixed. It turns out that a characteristic of viscosity-dominated flow is that it is impossible to set it into this kind of rotational motion. Mixing becomes difficult, even impossible.

On the other hand, the small scale of microfluidic channels and reactors can be exploited too. We can use the importance of surface energy to overcome the mixing problem. If we pattern the channel with chemical groups so that first one, then the other of the two fluids we wish to mix wet the surface, then as the two fluids are forced down the channel they will be continuously swapping positions. Or, rather than trying to overcome the fact that fluids in microfluidic flows do not mix, we can exploit it. If the two fluids react with each other, then in a microfluidic flow the reaction will be confined to the narrow region where the two fluids mix. This has been used to make very thin metal wires, for example.

How about controlling the flow of fluids in these microfluidic channels? One way exploits the properties of the responsive gels we introduced earlier. If the channel was lined with a thin layer of a polyacid gel, for example, then as long as the fluid flowing through the gel was acidic the layer would stay collapsed and the fluid would be able to flow freely. But if the acid is replaced by a base, then the gel would expand and choke off the channel—we have a valve that automatically responds to chemical changes in the fluid flowing through it.

The attractive feature of this application of responsive polymers is that the smaller the channel, the better it should operate. The volume change in a macroscopic piece of gel reflects changes in shape of the individual molecules of which it is composed. So if we can make a channel small enough, and if we can line it with molecules that change their shape in response to some stimulus, then we might be able to change the diameter of the channel enough to stop a molecule that was previously able to pass through. This would be, in effect, a switchable molecular valve.

If you make coffee using a filter machine, then you rely on the fact that filter paper has pores which let water pass though, but which are too small to permit the passage of the coffee grounds. The pores of common filter papers are tens of microns big, but it is possible to make a membrane whose pores are so small that even molecules only a little big larger than the water molecules themselves cannot pass through. Such a membrane can be used, for example, to remove salt from sea-water to yield drinkable fresh water. The sorting process is not free, though. To go from a high-entropy mixed state to a low-entropy, less-mixed state, energy has to be put into the system, and the water needs to forced through the membrane at high pressure.

At the low-technology end of this kind of nanoporous membrane is the common packaging material cellophane. This is made from cellulose derived from wood pulp, dissolved and then cast into a film in a process which naturally creates tiny pores through the material when it is made wet. A more high-technology product is the remarkable material made from track-etched polycarbonate (Nucleopore membrane). This starts life as a thin, completely impervious film of this rather robust plastic. These films are exposed to radiation; each high-energy particle travelling through the film produces a little, straight cylinder of damaged material which can be dissolved away, to yield a membrane traversed by cylindrical holes which can be as small as a few nanometres in diameter.

Given a sheet of material penetrated by nanoscale-sized pores, it is a simple matter in principle to make a membrane that responds to its environment, letting molecules through in one state, while blocking all traffic in another. If, for example, we lined the pore with a single molecular layer of a weak polybase, then, in acidic conditions, the polymer would ionise and swell up to fill the pore. In basic conditions, the un-ionised polymer would collapse onto the pore's walls, leaving a free passage down the centre for molecules to pass.

Controlling the motion of molecules through membranes is a central operation for all living things, and in biology we find responsive pores in which the principle of using a molecule that changes its shape in response to its environment to control the passage of molecules is refined to an extraordinarily sophisticated degree.

In the last chapter, we saw how membrane proteins were designed to self-assemble into position within a lipid membrane. Many of these proteins form pores, and it is conformational changes in the parts of the protein that form the walls of the pores that allow them to control their width, and thus the ease with which molecules of different shapes can get through. Among the many functions of biology that exploit this kind of channel, perhaps the most specialised are to be found in the nervous system. We will see in the next chapter that the walls of nerve cells are packed with channels that allow the selective transport of different kinds of ions. These channels can be opened or shut by various stimuli, the most important of which is the size of the electrical potential across the membrane. It is the operation of these so-called gated ion channels that forms the basis of the way our nerves generate and transmit their signals, allowing us to see and feel and think. We will discuss this in more detail in the next chapter, but first let us ask how different external signals can be used to control the shape of molecules. How can a soft machine measure and interpret its environment?

Sensors and transducers

What makes a robot different from a computer is that it can interact directly with its environment. To do this, it needs sensors and transducers. A sensor measures some aspect of the environment—whether it is light or dark, what the temperature is, what chemicals are in the atmosphere, for example. The robots we are making now are controlled by computers, so the sensors we use for robots tend to convert the information they receive about the environment into electrical signals, which can be converted into digital code for use as input for the computer that controls the robot. A transducer is a special kind of sensor which works both ways—it can convert information about the environment into an electrical signal, but it can also be driven by the electrical signal to modify the environment. The same basic mechanism that is used by a microphone to convert sound into electrical signals is used by a loudspeaker, which converts electrical signals into sound.

What will control a nanoscale robot? If this is to be a miniaturised electronic computer, constructed perhaps using the ideas of molecular electronics to be discussed in Chapter 8, then we will need to miniaturise the same kind of sensors and transducers that we use in big robots, so that all information about the environment can be converted into electrical signals for analysis. But there is another way—the way which is used by biology. Biology at the nanoscale processes information using chemical signals as well as electrical ones, and the basic mechanism it uses is, once again, the way that macromolecules can change their shape when their environment changes.

In large-scale robotics, converting electrical signals into motion is fairly fundamental. A large robot might do this using a hydraulic system, in a medium-sized robot it might be more convenient (particularly if rather precise motion is needed) to drive the motion directly with an electric motor. But at the interface between the macroscopic and microscopic world there is one very convenient and important mechanism that is worth talking about in more detail. This is the property of certain crystals to change their shape and size when a voltage is applied across them. The reverse property is also present; when these crystals are compressed, a voltage appears across the two faces that are being pressed together. This property is known as piezo-electricity.

To make a piezo-electric crystal, you need a material made up of units with a particular arrangement of positive and negative charges. Because positive and negative charges attract each other so strongly, a macroscopic lump of material, if left to itself, will always tend to have the charges exactly in balance overall. But it is possible for the balance between the charges to be slightly altered when the material is distorted, and this is the basis of piezo-electricity. To see how this works, think of a molecule which has a positively-charged part at one end and a negatively-charged part at the other—there is said to be a dipole. You can imagine this as being like a bar magnet, with the positive and negative parts of the molecule being like a north pole and a south pole. In a crystal, these bar magnets might be stacked in pairs, with the north pole of one magnet next to the south pole of its neighbour, so that the field is cancelled out. There are two ways of having a piezo-electric crystal. In the first way, the crystal is arranged in such a way that if the structure is distorted then the pairs of bar magnets are pushed slightly out of alignment. This means that the charges do not exactly cancel any more, and the crystal ends up with one surface slightly positively charged and one surface slightly negatively charged. This difference in charge between the two surfaces means that there is a voltage between them. In another type of piezo-electric material, even in the absence of any distortion of the crystal, pairs of dipoles do not quite cancel, and this mismatch is increased when the crystal is distorted.

So, in a piezo-electric material, if we distort the crystal then a voltage appears between opposite faces. We could use this voltage to measure the pressure that has been applied to the crystal. But the reverse phenomenon happens too; if we apply a voltage to the crystal its shape slightly changes. It is this voltage-induced shape change that makes piezo-electric materials so useful in nanotechnology.

The most common piezo-electric material is quartz—the crystalline version of silicon dioxide. The piezo-electric properties of quartz are what makes quartz crystal watches tell the time so accurately. Just as a wine glass makes a musical sound when you rub its rim, a quartz crystal will have a resonant frequency; the piezo-electricity of the quartz means that the mechanical resonance that makes it produce a single tone of sound also produces an electrical signal which oscillates at a precise frequency. Just as a grandfather clock tells the time by counting the periodic swings of a pendulum, a digital watch counts the oscillations of its quartz crystal. The oscillations of the crystal, though, are much higher in frequency than the swinging of the pendulum, and much more precise in their timekeeping.

Another useful piezo-electric material is the polymer poly(vinylidene fluoride). This is easily made into thin, flexible sheets, very similar in appearance to sheets of polythene. The material lends itself to making large-area, flat microphones and loudspeakers.

But the unsung piezo-electric star of nanotechnology is an obscure-sounding ceramic, lead zirconium titanate. This material is universally known as PZT, and it is essential for even British scientists to pronounce the Z American-style, to get the rhyme. What one can do with a block of PZT is to use it as a probe that can make nanometre-scale movements in response to precisely-controlled voltages.

One of the earliest pieces of scientific equipment to exploit this fully was a device made by David Tabor and his then student Jacob Israelachvili in the Cavendish Laboratory in Cambridge. This was a force measuring device—essentially a spring balance—that could measure the forces between surfaces as they approached toward molecular contact. Two surfaces that were flat on atomic scales were positioned a few nanometres apart, their separation being precisely controlled by piezo-electric crystals. One of the surfaces was mounted on a very weak spring, and, as the two surfaces pushed each other apart or pulled each other together, this force could be measured by detecting how much the spring was bent.

It is this ability to move surfaces and probes around with nanometre control that underlies the operation of scanning probe microscopes, like the scanning tunnelling microscope and the atomic force microscope that we talked about in Chapter 2. By joining pairs of PZT crystals, one can get a device that slightly bends when a voltage is applied. This is how the probe of an atomic force microscope or scanning tunnelling microscope is moved around with sufficient precision to pick up individual atoms or molecules.

Piezoelectric materials produce a nanoscale change of shape in a macroscopic piece of material: the effect is too subtle to be useful for wholly nanoscale devices. Voltage is also the trigger for a mechanical change in one important class of biological nano-machine—the voltage-gated ion channels that underlie the operation of our nerves and brains. These pores in the lipid membrane are made from proteins that pass completely through the membrane. Very roughly speaking, one can think of the structure of the ion channel as being made up of four cylinders arranged in a square in the plane of the membrane. The gap in the middle of the square is the pore. The surfaces of the cylinders are knobbly, rather than smooth, and the knobbles will come together at one point in the pore to make a pinch point, a narrowing that ensures that only one kind of ion can easily get through the gap. Of course, the cylinders are not rigid; they are soft self-assembled macromolecules which, on a local scale, are wobbling around in response to Brownian motion.

If a voltage is applied across the membrane, then within the membrane there will be an electric field. Because the thickness of the membrane is so small, it does not take very much of a voltage to generate quite a large field. Since proteins have charged groups, when the field is turned on the positive charges will be pulled one way across the membrane and the negative charges the other way. This can slightly destabilise the structure of the proteins, pulling it into a slightly different shape, which pinches off the pore and closes the gate.

Using a conformational change to open and close a gate in the pore of a membrane is a very common trick in biology. In many other cases, it is the binding of a specific messenger molecule to the channel that causes the conformational change which opens or shuts it. We will see in the next chapter that this trick is the basis, not only of the way biology controls the traffic of molecules around organisms, but also of the way that it processes information.

7

Wetware: chemical computing from bacteria to brains

———⊃∘○∘⊂———

Introduction—Galvani and the chemical computer

The connection between electricity and the way our brains and nervous systems work was made long ago. The Italian scientist Galvani famously made a dissected frog's leg twitch by connecting a battery to it, and the idea that you can somehow jump-start a nervous system back into life by applying electricity to it irreversibly entered popular western culture in Mary Shelley's *Frankenstein*. Thus, when the first electronic calculating machines—computers—were developed in the 1940s it was only too natural to think of them as analogues to our own nervous systems—'electronic brains'. As computers have become embedded in our everyday lives, it becomes even more difficult to avoid making analogies between the brain and computers. Yet these analogies are profoundly misleading. Yes, the action of both brains and computers both involve electricity in some way. But whereas, as we have seen, the action of a computer is based on the transport of electrons through materials, in nervous systems what moves are not electrons, but ions—whole, charged atoms and molecules. The mechanism by which these ions move around in their watery environment is nothing to do with the intrinsically quantum mechanical movement of electrons in semiconductors. Instead, at the heart of these mechanisms are the now-familiar companions of the wet nanoworld, Brownian motion and conformational change. The basic elements of logic in living creatures are molecules, and the way information is moved around is not by the ghostly quantum mechanisms underlying electronic transport, but by the very physical movement of molecules, driven by that ever-present Brownian motion.

Using molecules to communicate, signal, and compute is a very natural design decision if you are operating in a warm, wet environment on the nanoscale. We saw, when we talked about Brownian motion and diffusion in Chapter 4, that on the length scale of a bacterium Brownian motion is highly efficient at moving molecules around. If a molecule is created at one point in the cell, then it will end up being very rapidly moved to the signal's destination, driven only by the random buffeting of the water molecules. But we saw that the way the efficiency of diffusion scales with size makes it very unfavourable to use the same principles when you get large, even when you have ways of helping diffusion along, like blood circulation systems. Big organisms like humans do have a remnant of the simple diffusion-based ways of carrying messages—our endocrine system, in which hormones are released into the blood system in one part of our body, and signal to other parts of our bodies to make changes in response. So, during our 'fight or flight' response, adrenaline is released by one organ and the molecules make their way to all parts of our body, putting it into a state of readiness for sudden physical activity. But this kind of communication is far too unspecific to carry the huge variety of specific messages a big animal like ourselves needs to run our lives. We need to know that we have dropped a brick on our foot, when we can smell gas, or that a tiger is coming towards us. To carry all these signals and to do computations based on them, we have evolved a nervous system.

The nervous system is also based on the diffusion of various types of molecules; the controlled motion of small ions backwards and forwards across the membranes of nerve cells is the way in which electrical signals are transmitted down the long protrusions that make up neurons and axons, while larger signalling molecules convey the signals across the gaps—the synapses.

Reflex, instinct, and intelligence

As humans, we are all agreed that we can, in principle, display intelligence. Faced with a problem that needs to be overcome before we can achieve some desired goal, we are able to analyse it in terms of sets of rules or models that we have reason to believe describe the domain in which the problem is found, and identify a course of action that will allow us to solve the problem and achieve the goal. We may perhaps allow that some other animals—dolphins, apes, dogs, and cats, perhaps—also show some measure of intelligence, but for so-called lower animals we are more comfortable talking about behaviours that are governed by instinct. By this we mean a set of rather mechanically programmed responses to external stimuli. Nonetheless, the types of computations that organisms must undertake to produce what we refer to as instinctive behaviour can be very complex—think of the way fish like salmon or eels can find their way back over huge distances to their spawning grounds, or the way honey-bees, by their 'dance', can communicate information about the location of sources of nectar. The simplest kinds of information processing in living things we refer to as a reflex. If our hand comes into contact with a hot surface then we will pull it away by reflex; the signal from the burned hand produces a signal to the arm muscle to withdraw the hand without needing to go to the conscious or problem-solving parts of our brain. Even the simplest organisms usually seem to have reflexes, and, as we will see, to have a reflex it is not even necessary to have a nervous system of any kind. Plants can turn their leaves toward the light, and even a bacterium can swim toward a food source.

In the material world, we can design control systems that display behaviours that have something in common with the reflexive or instinctive behaviour of animals. A home central heating system might have a thermostat; when the temperature falls below the set point a relay switches the heating system on, allowing one's home to adapt in this simple way to changes in the environment. Computer scientists have long talked about intelligent systems—computers whose responses to inputs mimic, in some very restricted domain, the responses that an informed human might make. Engineers and materials scientists even talk about intelligent or smart structures or materials. These are materials that are able to sense some aspect of the environment around them, and respond in a way that adapts to this change.

There is an obvious difficulty in using the word intelligence to cover this kind of adaptive, but essentially inanimate, behaviour. We find it difficult to define intelligence without referring to essentially human attributes; the most convincing test that has been devised to find out whether a computer program is truly intelligent was proposed by the computer science pioneer, Alan Turing. Imagine a human tester isolated in a room, able to communicate with the outside world only through e-mail. We introduce the tester to a new correspondent, with whom he can have a conversation through e-mail on any subject he chooses. We hear a lot nowadays of people who meet only through e-mail, and indeed form close relationships without ever physically meeting; we also hear all too often about people who exploit the fact that, in e-mail interactions, it is all too easy to pretend to be something one is not. An unattractive and unsuccessful man might pretend to be a rich and eligible suitor in an attempt to find a girlfriend. But what if the new e-mail correspondent we have introduced to our human tester is not a human at all, but simply a sophisticated computer program? If the program did manage to fool the tester into believing he was interacting with another human being, then we would certainly be justified in saying that that computer program had real intelligence.

The Turing test is very difficult—surprisingly so, to those who have not thought through just how much background and contextual knowledge underlies even quite simple conversations. No computer program has yet come near to passing a Turing test that has not been massively rigged in its favour, so it is difficult to argue that any machine yet possesses intelligence in this sense. Obviously, plants, bacteria, worms, or even dolphins are not going to pass the test either.[21] Nonetheless, I need a word to describe the kind of integrated, adaptive response to stimuli that goes beyond the simplest kind of reflex, and for this I cannot think of anything better than to continue to use the word 'intelligent', all the while remembering that I am using it in this restricted sense.

We can now distinguish between reflexive behaviour and intelligent behaviour. A polymer gel may behave in a reflexive way; if we heat it up, it

[21] To be fair, one should stress that Turing himself did not consider the test to be an exclusive one; while a machine that passed the test could certainly be said to be intelligent, one cannot automatically conclude that a machine that failed the test is not intelligent.

shrinks, and if we cool it down, it expands again. We would anticipate that intelligent behaviour requires more than this. One of the basic ingredients of intelligence must be the ability to do logical computations—to evaluate rules which specify how combinations of input combine to determine different outcomes. A simple thermostat has one rule governing its behaviour—if the temperature is above the set point, turn the boiler off; if it is below the set point, turn it on. Actually, this kind of thermostat does not work well if you need to control the temperature very closely; usually what you end up with is a temperature that oscillates around the set point without converging on it. If it is important to keep the temperature very close to the set point, for example in the sort of incubator you use to keep a highly-premature baby alive, then you would need to use a more sophisticated temperature controller, that has a bigger range of responses than simply turning the heater on or off. The controller would measure how much the temperature differed from the set point, and it would set the heater output in proportion to this difference. So if the temperature is a long way below the set point then the heater is turned up high, while as the temperature gets closer to the set point the heater setting is lowered. Because it takes time for heat to spread from the heater to all of the distant parts of the incubator, we need to build in a lag time; the incubator will carry on heating up even after the heater is switched off. So we need to calculate, not just how different the temperature is from the set-point temperature, but how quickly that difference is changing, and we need to change the setting of the heater in response to that information as well.

We would need to program this kind of temperature controller—we need to tell it in advance what weight it needs to give the various pieces of information, like how big the difference is between the temperature and the setpoint, and how fast the temperature is changing, when it decides how high the heater needs to be set. The details of the program will depend on the nature of the system you are trying to control—the power output of the heater, the heat capacity of the incubator, how good the thermal insulation is, and so on. Some fairly complex mathematical theory allows you to calculate some predicted settings, given information about your system. I have to admit, though, that as an unsophisticated experimental physicist, when I have had to connect up a home-built oven to this kind of temperature controller, I have tended to mess around with the settings more or less by trial and error until the thing seems to work. But for a little bit more money you can buy a controller that, when you wire it up to your home-made heater and oven, will do this work for you—by testing how the temperature changes as it turns the heater on, it will itself compute its own optimum settings.

This is a higher level of adaptive behaviour. We have a fairly sophisticated program that determines what response the device should make to a given stimulus, taking into account, not only the current value of the stimulus, but also some degree of history. But, in addition to this, the system can also adapt its own program to take account of its surroundings. In this sense, the controller has a similar level of intelligence to that which we use when we do something like throw a ball. Not only does our brain compute how much each of our arm and shoulder muscles need to be stimulated in order to fling the missile the desired distance in the right direction, we are also able to adapt the calculation in response to our judgement of how heavy the ball is.

It is at least this level of intelligence that we are going to need to achieve if we are going to make nanoscale machines that can do useful things, like deliver drugs or repair tissue. On the macroscopic scale, such as for temperature controllers, such control systems are now almost universally based on microprocessors—effectively small, specialised computers based on silicon integrated circuits. But on the nanoscale, organisms like bacteria exhibit similar levels of adaptive control. This is based, not on electronics, but on logic carried out by conformational changes in large molecules, and on information being carried by the diffusion of molecules. We will see how this works in the next section.

How E. Coli responds to its environment

Living organisms need to eat. Life needs a continual input of energy and raw materials; the way this input is achieved is the most important factor controlling the design of the organism as a whole. One fundamental distinction is between organisms which get their energy from sunlight, and having found a light spot to grow in can stay put, and organisms which have to move around to find high-energy chemicals to feed on. At the macroscopic level, we have the distinction between plants and animals, with plants manufacturing energy-rich sugars from sunlight, water, and carbon dioxide, and animals either eating the plants or eating other animals that themselves eat plants, thus also ultimately getting their sustenance from plants, but at one or more removes.

The distinction between plants and animals does not work so well for single-celled creatures, and now we regard the fundamental division as being between prokaryotes—bacteria and their allies—which are made up of simple cells with little internal structure, and eukaryotes, which may be single- or multi-cellular, but whose cells have a complex internal structure. Eukaryotes include plants, animals, and fungi. Both fundamental classes of organisms— prokaryotes and eukaryotes—each include some organisms that get their energy directly from sunlight, and some that have to move around in search of food.

One of the most well-known types of bacteria—the E. Coli that inhabit the guts of humans and other animals—has a very well-developed system that allows them to move toward sources of food (and away from sources of harmful chemicals). This system is now quite well understood, and gives us a very good example of the way in which the smallest and apparently simplest organisms can process and act on information in a surprisingly sophisticated way.

We saw in Chapter 6 that bacteria like E. Coli are equipped with rotary motors embedded in their cell walls, which, driven by the energy stored in a gradient of hydrogen ions, can turn a whip-like thread or flagellum, thus propelling the bacterium forward. Also present in the cell wall of the bacteria are sensors, which are able to detect a source of food molecules. Signals from the sensors are transmitted to the motors, which modulate their operation in such a way that the bacteria moves toward the food source.

Stated in this way, we can imagine how we might build a robot submarine that had some similar property. Chemical sensors would detect the food chemicals, producing an electrical output proportional to the concentration of the chemicals. These sensor outputs would be converted from analogue signals to digital signals, which would be input into a microprocessor. The microprocessor would be programmed to produce outputs that would control the motor driving the submarine and the steering device controlling its direction, in such a way that the robot was always kept on a heading that led it closer to the food.

The way in which the control system of a bacterium operates has some superficial similarities with our hypothetical robot, but it also has some fascinating differences that illustrate graphically how the nanoworld differs from the macroworld. Firstly, the bacteria has no steering device, no rudder—not surprisingly, as it is quite difficult to build a rudder that works in the conditions dominated by viscosity that characterise the nanoworld. Instead, its rotary motors have only two controls. The two settings available to the motor correspond to its rotation in the two opposite senses available to it. Because the filament that the motor drives has an intrinsic handedness, the two settings produce quite different behaviours. If the motors are running in a counterclockwise direction, then the filaments from adjacent motors are drawn together, and the effect is that they all work to propel the bacterium in the same direction. If the motors switch direction to rotate in a clockwise sense, then the filaments fly apart, and the bacterium randomly tumble in direction.

In normal operation, the two types of motion alternate. The bacterium alternates between periods of swimming in a straight line and periods of random tumbling, producing random changes in direction. In this mode, the bacterium makes progress by a random walk. So, lacking a rudder, this kind of bacterium is somewhat like a novice skier who has not yet learnt to turn, so that if he needs to change direction he has to fall over and then get up again, facing a different way.

The effect of the sensors is to change the frequency with which the motors change direction. If the concentration of food is increasing as the bacterium swims, then the switch of direction is delayed, and the period of swimming straight, in the direction of the food, is prolonged. The net result is that, what was, in the absence of any signs of food, a pure random walk, is biased so that steps toward the food source are favoured. How is the information from the sensors carried to the motors? Rather than the wires that would carry the messages in our robot submarines, the bacterium uses chemical messengers.

The sensor itself consists of a pair of closely-associated protein molecules that spans the cell membrane of the bacterium. On the outer surface of the sensor, there is a slot with the right pattern of sticky patches to make a strong bond with food chemicals, for example sugars, or with toxic or repellent chemicals. When a toxic molecule encounters this binding site, it initiates a series of events that lead to a chemically-activated messenger molecule being released. This messenger diffuses through the interior of the bacterium, driven only by the random buffeting of the water molecules around it. If chance brings it to one of the motors, then it binds to another receptor and sets in train the changes that cause the motor to change direction; and this leads the bacterium to stop swimming straight, and to start to tumble. But the lifetime of its message is limited; if the molecule has not reached its destination within about ten seconds then it is deactivated.

So, rather than the wires that we would use in a macroscopic robot, in bacteria the signals are carried by diffusing molecules. Where in the bacteria is the equivalent of the microprocessor? The surprising answer is that it is the combination of the sensor molecules themselves and the environment of the cell that carries out the information processing—in effect, the sensor is a component of a molecular computer. If a food molecule binds to the sensor, then the chain of events leading to the release of the messenger molecule is stopped. What commands arrive at the motors depends on the balance of signals sent out by the sensors. In a high concentration of food molecules, most of the receptors will be inhibited and few messenger molecules will be sent out. The probability of the bacterium changing from swimming to tumbling is relatively small. But, if there are few food molecules around, and relatively high concentrations of toxic, repellent chemicals, then many messengers will be dispatched, and there will be a high probability that enough will reach the motors to cause the bacterium to change direction.

As described so far, the control system seems not too dissimilar to the simple central heating thermostat, with a simple two-state response. The program seems fairly simple, and can be stated in words something like this: add up the total number of detections of food molecules and subtract the number of detections of toxins; if the resulting score is positive, then keep going straight, if it is negative, then tumble. The only thing that would make it a little different from a thermostat is the fact that, rather than there being a simple on/off switch, the motor control is probabilistic; the higher the concentration of messengers in the bacterium, the bigger the chance of the motors changing direction.

Two extra factors make the behaviour of the bacterium's control system considerably richer. Firstly, the more the sensors are exposed to their target molecules, the less sensitive they become. This means that, like our more sophisticated temperature controller, the bacterium's motion control system is not just sensitive to whether there are more food molecules than toxins in the vicinity, it detects the rate at which the concentrations are changing. In order to keep on swimming straight, the bacterium needs to be travelling toward the source of the food. Continuously increasing concentrations of food molecules are needed to suppress the command to the motors to change direction and start tumbling.

The other complication is that the rate at which the messenger molecules are deactivated is itself controlled by the presence of another protein, which catalyses the deactivation reaction. This gives the bacterium another mechanism to modulate the control process; if there was some other reason for which the bacterium might want to decrease or increase the frequency of tumbling, then it can do this by increasing or decreasing the concentration of the catalyst molecule. One can think of this as an extra communication port by which the output of another control module can be input into the motor control module. It is as if, in our temperature controller, we have an additional plug, so that if we wanted to include another factor into the calculation of whether to turn the heater on, then we could.

We have discussed the motion control system in E. Coli in some detail, partly because it is by far the best understood control system in a single-celled organism, and partly because it is fairly simple and easy to visualise in terms of the types of man-made control system we know from the macroscopic world. But it also turns out to involve pretty much all of the general principles which underlie communication and logic, both within cells and between cells in both single-celled and multi-celled organisms. Informed by this specific example, let us take a look at some of the general principles that nature seems to use in chemical computing.

The principles of chemical computing

Cells of all kinds do their information processing by chemical means; information is carried from place to place within the cell, and between cells too, by the use of messenger molecules rather than by wires. This scheme has a number of advantages. We do not need to build any particularly complicated infrastructure to link the places where the messages are generated to the places which receive them, because we can rely on Brownian motion to diffuse the messages all over the cell. We can have many messages being carried simultaneously. The disadvantage might seem to be that messages might be rather slow to arrive, if we have to wait for the messenger molecule to randomly shuffle around until it finds its destination. But, on the other hand, in Chapter 4 we saw how much more efficient diffusion was at transporting things over small distances compared to big ones.

A bacterium, then, does not have a brain, but it does have an immensely sophisticated information processing system. It is based on the diffusion of chemicals, driven by Brownian motion, to move information from place to place, and on conformational changes in macromolecules in response to molecular recognition of these information-carrying chemicals. This is a system of immense complexity, which we are barely beginning to understand. At the cell level, this chemical computer operates in all organisms, including complex multi-cellular ones like ourselves. Large animals like ourselves need communication and information processing on a larger scale as well, on the scale of the organism rather than the cell. Some of this communication is also performed by the use of signalling molecules. As we have already mentioned, when we detect imminent danger, the hormone adrenaline is released into our bloodstream, causing our heart rate to increase, and our muscles to break down glycogen in preparation for energetic action. On a slower time-scale, hormones released during puberty seem to drive teenagers to paint their bedrooms purple and listen to dismal music. But chemical signalling of this simple kind has drawbacks for larger organisms, in terms of a rather slow rate of response, and we have evolved another communication and information processing system, the brain and nervous system. This, at a superficial level, looks much more like an electronic computer, but in reality it relies on the same principles as the simpler chemical computers of single cells—diffusion and conformational changes in macromolecules.

How can a single molecule act as a logic element in a computer? The answer, yet again, lies in the fact that the self-assembled structure of a protein can change its shape in response to a change in the environment. We have already met this principle when we talked about molecular motors; there, protein molecules adopt two different shapes according to whether they are in the presence of ATP or ADP. For a motor protein, we exploit the change in shape of the molecule to produce mechanical work. But what about the other functions of a protein, for example its ability to catalyse a chemical reaction? This ability depends very strongly on the shape the protein takes up; so, if you can change that shape by the presence of some other molecule, then you can switch the reaction that the protein catalyses on or off according to whether that molecule is present or not. You can think of this process as an elementary logical operation.

Let us call the molecule that causes the change of shape A, and the output of the reaction that the protein catalyses B. Here A is the input to the logical operation, whose output is B. We can express the logical operation as if(A)then(B) else(notB). Different components can be linked together; if we have another protein whose activity is switched on or off by B, then the output of this protein (call it C) depends on whether A is present by an indirect route. We can imagine more complicated logical steps, too—a protein might need two inputs—D and E—in order to catalyse the production of product F; now we have the operation if(D and E)then(F)else(notF). Even more complexity is introduced by the fact that responses can be dependent on history; the sensitivity of the protein to the input molecule A may be reduced with increasing exposure. It is the combination of all of these kinds of process that is known to biologists as cell signalling.

The details of these processes are dauntingly complex, and it is easy to get bogged down in the details of G-proteins, phosphorylation cascades, cyclic-AMP-dependent protein kinase, and the rest. The situation is made worse by the fact that biologists in this field seem to have completely lost their grip when it comes to nomenclature. Signalling proteins rejoice in names like Groucho, Hedgehog, and Dishevelled, which makes reading the literature in this area a rather surreal experience. But on top of the complexity of the individual processes is a complexity at a systems level which is barely understood at all.

The social life of cells

A society is built on communication. A multi-cellular organism, such as a human being, can be thought of as a closely-integrated community of genetically-identical cells, which adopt different forms according to their function and usefulness for the organism as a whole. The development and smooth operation of the organism depends on constant communication between its constituent cells. Each of the steps that led to truly multi-cellular organisms depended on the development and refinement of chemical signalling between different cells.

Even bacteria can be surprisingly sociable beings. Many bacteria can switch between different states on receipt of messages from other nearby bacteria; the messenger is usually a simple derivative of ATP called cyclic AMP. The most highly-developed social lifestyle is shared both by a class of bacteria (myxobacteria) and by some amoeba, which are single-celled eukaryotes. Best known among these are the slime moulds, typically found on rotting logs. For much of their life these creatures live their lives as single-celled, free-living amoeba, moving around using their ability to push out a protrusion of their cell wall—a pseudopod—and then ooze into it. For food, they graze on what bacteria they can find. But when times get hard, and the food supply falls and there are too many amoeba for the supply to feed, they start to release cyclic AMP. The response of the amoeba to this collective distress call is to stop eating, to stop dividing, and to gather together for safety.

Stimulated by cyclic AMP, the amoeba come together in a dense mass of cells, which coats itself with slime. This slug-like body can then wander off as a unit, searching for somewhere a bit drier. When it finds a dry spot, it changes shape, forming a bud-like structure which is attached to the ground by a long stalk. This releases spores—individual cells wrapped in a tough outer coating— which are dispersed to other locations. If a spore finds a damp environment then an amoeba will emerge from the tough outer skin and the whole cycle will begin again.

Slime moulds are examples of single-celled organisms that have taken the first steps on the road to becoming true multi-celled organisms such as plants and animals, and this first step depends on chemical signalling between different cells. Animals are distinguished by the fact that they develop from an embryo formed by the fertilization of an egg by a sperm. Every cell that forms from the original fertilized egg is genetically identical—yet somehow, in the process by which the embryo grows, the cells differentiate to take on very different roles. Some turn into skin cells, some make muscle and bone, and others turn into the very specialised cells that the nervous system is made of—neurons.

What tells something that starts out as an undifferentiated cell in the embryo of an animal—a stem cell—to transform itself into one of these highly-differentiated and specialised cells? Usually, it is the influence of a signalling molecule, which controls which of the genes are put into operation and which are prevented from being expressed. The details of animal development are fascinating as well as dauntingly complex, but at the bottom of it all are sequences of events initiated by the change in shape of proteins when another protein binds to them.

Why big animals needed to develop a longer-ranged signalling mechanism

Chemical signalling is fine for bacteria, and it has been evolved to a peak of perfection in the way it controls the development of multi-cellular animals. Animals still use chemicals to carry information around the body, in the form of hormones like the familiar adrenaline (also known as epinephrene), which prepares the body for physical exertion when danger is detected by increasing the heart rate and promoting glycogen breakdown in the muscles. But this kind of general broadcasting of information—using what is called the endocrine system—has its limitations; it is not very specific, and the range of different messages that can be carried is rather limited.

For this reason, animals evolved a new way of carrying information. Rather than broadcasting the message throughout the body, this method carries it point to point. This new information route is the nervous system. This is made up of the highly-specialised cells called neurons, which are distinguished from other cells by their size and their form. A neuron consists of a cell body from which highly-branched dendrites protrude (see figure 7.1). In addition to these dendrites, there is usually one protrusion which is

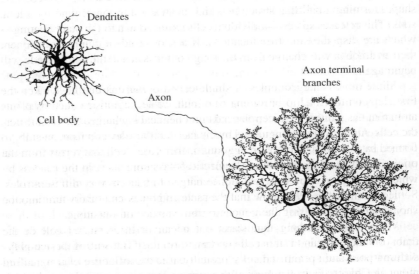

Dendrites

Axon terminal branches

Axon

Cell body

Fig. 7.1 Typical vertebrate neuron. Reproduced with permission from B. B. Boycott. In *Essays on the nervous system* (ed. R. Bellairs and E. G. Gray). Clarendon Press, Oxford, 1974.

unbranched and very long—up to one metre in length in a vertebrate like a human. It is along this *axon* that nerve signals are carried over long distances. The end of the axon also branches, and these branches make specialised junctions with the cell bodies and dendrites of other nerve cells. These junctions are called synapses. The extreme degree of branching, both of the dendrites and the terminals of the axon, mean that a single neuron will interact with a very large number of other neurons—as many as 100 000.

Neurons process information in two ways—signals travel within neurons, down the axon, in the form of electrical pulses, and signals are passed between neurons at the synapses, where the transmission of signalling molecules causes channels in the membrane to open or close. But, despite the apparent differences between the two mechanisms, it is the same basic principle—the change in shape of a protein in response to a change in its environment—that underlies them both.

Nervous energy

Brainwork takes a lot of energy, we are sometimes told, but it is difficult to see why. Athletic types scorn the idea that sitting around thinking is even in the same league as more obviously energetic pursuits like kicking balls or running up and down pitches. But the sportsmen are wrong; our nervous systems use a prodigious amount of energy, even though at the macroscopic level there does not seem to be very much to show for it physically.

Everyone knows that bodily fluids—tears, sweat, blood, and the rest—are salty. It is often said that this reflects the salt-water environment of the sea in which the first animals evolved. But the composition within the cells of animals is actually very different from either sea-water or from the fluids that they are surrounded by. The chemical name for common salt is sodium chloride, but when salt is dissolved in water the sodium part and the chloride part live quite separate lives, as charged ions. A sodium ion is a sodium atom that has lost one electron, giving it a net positive charge of one, while a chloride ion is a chlorine atom that has gained one electron, making it negatively charged. Inside the cell, the concentration of sodium ions is much less than the concentration outside. To compensate for this deficiency in sodium, there are very many more ions of potassium inside the cell—both sodium and potassium ions carrying a net positive charge of one (in units in which the charge of an electron is minus one).

How can this be? We know that the random processes of Brownian motion should end up equalising the concentrations inside and outside each of these cells. To work against this randomising tendency there must be some device that sorts the ions, and it must use energy to do this. The sorting device is the sodium–potassium pump. This is a protein-based motor of the kind we talked about in Chapter 6; it is a molecular machine that sits across the membrane, and binds sodium ions from inside the membrane and potassium ions from outside the membrane. Driven by the chemical fuel of ATP, the machine pumps the sodium out of the cell and the potassium into it. A continuous supply of energy is needed to do this; the sodium–potassium pump is fighting the effect of entropy by sorting the two types of ions between the inside and outside of the cell. Just as it takes a continual input of energy to keep a child's bedroom tidy, so some types of cell can use up as much as two-thirds of their total energy input in keeping the sodium and potassium properly sorted. For the brain, half of the energy it uses is taken up by these pumps.

Part of the reason for the amount of energy it takes to keep the inside of the cell rich in potassium is that the cell membrane is made deliberately leaky to potassium. In the membrane there are special pores which are exactly the right size, and have exactly the right lining, so that they can let the potassium ions pass, but block the passage of sodium ions. So some potassium ions leak out from the inside of the cell to the outside. But, because the potassium ions are positively charged, this sets up an electrical field. Positive ions move from inside to outside, but because the pores are so selective there is nothing to balance this movement of positive charges from inside to outside the membrane. Sodium ions cannot move into the cell, nor can negative ions move out, so there ends up being a net difference in charge between the two sides of the membrane. The outside surface of the membrane ends up building up more and more positive charge, which repels the potassium ions until, in the end, no more cross the membrane. This difference in charge between the two surfaces of the membrane means that there is a voltage across the membrane, just as there is a voltage between the two terminals of a battery. This voltage is called the membrane potential.

Nerve cells, then, are kept in a high-energy state—what physiologists call an excitable state—waiting for something to happen. What it is they are waiting for is a pulse, a signal. The way the signal propagates is like this. The membrane is studded with voltage-gated ion channels which, when open, allow sodium to pass through—the sodium channels. If the voltage across the membrane falls for any reason, then the sodium channels open up. Sodium ions move into the cell, cancelling out the net negative charge inside the membrane. The potential across the membrane falls even lower. More sodium rushes in, reinforcing this drop in potential even more. Nearby sodium channels start to feel the reduction in potential, and they open up too. A shock wave of reduced membrane potential travels along the nerve axon.

What happens to the axon once the pulse of reduced membrane potential has gone by? There must be a mechanism to reset the system and bring the membrane potential back to its resting value. This happens in two ways. Firstly, the sodium-gated ion channels only stay open for a certain short time, even if the membrane potential stays small. Secondly, voltage-gated potassium channels open up after a short delay, so that more potassium can leak from inside the cell to the outside. The result is that the membrane potential readjusts to its normal, excitable state and the nerve is ready for another pulse to be transmitted.

The key thing to remember about a nerve pulse is that, although it manifests itself as a pulse of electrical potential, it is not the movement of electrons that is carrying the information, but ultimately the movement of ions through the membrane. This makes nerves very much slower than electrical cables to transmit a signal. The signal that carries your television picture from the aerial to the television set travels at the speed of light (modified for the dielectric properties of the insulator in the coaxial cable)—a few hundred million metres per second. But a nerve transmits a signal at a very much more stately speed of between one and one hundred metres per second. There is quite a perceptible time delay between the detection of the fact that you have stubbed your toe and the arrival of that information at your brain.

If information is transmitted along the axon as a pulse of reduced membrane potential, then how is it transmitted between nerve cells? When the pulse arrives at the terminal branches of the axon, at the synapses where these branches form connections with other neurons, it causes the release of small signalling molecules called neurotransmitters. These molecules diffuse, by simple Brownian motion, across the gap, where they bind with receptor proteins. These receptors work in the increasingly familiar way—when the neurotransmitters bind to them, they change their shape, and this change of shape makes a channel through the membrane open. Different neurotransmitters open different channels. Some channels open to let sodium ions through, and this has the effect of making it easier to set off a new nerve pulse. Other channels open to let potassium or chlorine ions through, which has the effect of making it harder to set off a new pulse. Since a single nerve cell is receiving inputs from many other different nerve cells through different types of synapse, the net result is that the nerve cell is performing a complicated logic. According to the way in which all of the different inputs add up, some promoting firing of the nerve cell and some inhibiting firing, different combinations of input can lead to different outputs.

Now we can begin to understand how nervous systems can process information—how they can receive information from a variety of sources and produce an output that depends, by some rules, on the various combinations of input. But animals do more than make decisions on the basis of the information they receive at any given instant; they learn and remember. The rules that they use to decide what to do are influenced by what has happened in the past.

Human learning and memory is fantastically rich and complicated, and, despite the bravest attempts of experimental psychologists, it is too challenging a task to try to associate memory with any physical changes in the brain. But animals with simpler nervous systems can remember things, too, and perhaps the most insight has come from studying a coastal mollusc called the sea-hare. If you have been paddling in shallow waters on temperate coasts, then you may have seen these inconspicuous creatures—grazing on brown seaweed, a sea-hare itself looks like a broken-off scrap of seaweed at first sight, until you see that the scrap can move surprisingly briskly, revealing itself as a rather baroque slug.

A sea-hare is not an intellectual giant. It gets by with a nervous system of only about 20 000 neurons, compared with the two billion or so neurons in a human brain. But, even with this rather limited mental equipment, it is still capable of learning. Just as Pavlov famously conditioned dogs to salivate when he rang a bell, a sea-hare can be trained to contract its gill when a light is shone on it. But the nervous system of the sea-hare is simple enough that the molecular changes that underlie learning can be picked out.

The basic process underlying learning is a change in the way a synapse responds. This change can be permanent, the result being that after the change it takes a smaller stimulus at the synapse to cause the nerve to fire. One possible way in which this could happen is by another cell-signalling event, in which the binding of the neurotransmitter causes the production of the signalling molecule cyclic AMP, which in turn causes a permanent modification of the potassium channels. Remember that this is the same molecule that makes a slime mould aggregate; it is an interesting comment on the degree to which nature reuses a few basic themes to do more and more complicated things that this molecule is implicated in those most human of capabilities, memory and learning. But, on the other hand, I should stress that the question of how synapses remould themselves in response to experience is undoubtedly very complicated and very far from being understood completely.

How brains are different from computers

The temptation to make an analogy between the brain and a computer is so strong that it is worth explicitly listing some of the ways in which silicon logic and brain logic are quite different.

First of all, there is a pronounced difference in size and architecture. Brains may not have as many individual components as a supercomputer, but each of these components has many more connections (see figure 7.2). A brain has roughly twenty billion neurons, while a supercomputer might have one hundred billion transistors in its memory and central processing unit. But in the supercomputer, each transistor only has three connections. In the brain, though, the number of synapses—the connections between different neurons—is more than two hundred thousand billion.

Then, there is a fundamental difference between the digital logic used by the computer and the chemical logic used by the brain. Digital logic works absolutely accurately, absolutely reliably, absolutely reproducibly. If you put the same input into a computer many times, then you will always get the same answer out. The brain works differently. Because the way signals are transmitted across synapses depends on the Brownian motion of the neurotransmitters across the gap, there is an inescapable element of chance or randomness about synaptic logic. The operation of the synapses depends on history and environment as well; in particular, the overall way the brain works can be affected by the overall concentration of certain chemicals or neuromodulators. If I have drunk a bottle of wine then my thinking starts to get a bit blurred, but there is no way I can make my laptop drunk!

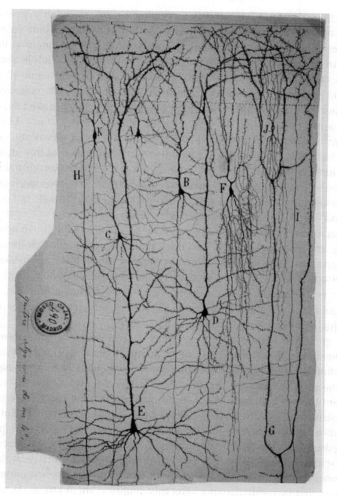

Fig. 7.2 Brain cells in the cerebrum of a one-month old child, as revealed by the Golgi staining technique, drawn by Ramón y Cajal in 1904. Figure courtesy of Cajal Institute, CSIC, Madrid, Spain, copyright inheritors of Santiago Ramón y Cajal.

The design principles of brains and computers are very different, too. The basic units of computers are identical; complexity arises simply from linking together very many of these identical units together. But in the brain, no two synapses are the same. In computers, there is a clear distinction between the adaptive part—the memory—and the logic processor—the central processing unit. This separation between memory and logic is quite unlike the brain, where adaptation and memory is intimately bound up with processing, as a result of the way the response of synapses can be changed with time.

But the most fundamental difference that I think everybody would agree on is that associated in some way with our brains is an attribute called consciousness, that we have but which computers do not. A vast amount has been written in search of a 'neural correlate of consciousness', which is a nice way of expressing the idea that science should be able to find some physically definable reality underlying the sensations that we know as consciousness. This is not the place to add to this literature, but I do want to make one, essentially negative, point.

Many writers have attempted to find some connection between quantum mechanics and consciousness. I do not really see the appeal of this idea, which I sometimes suspect is based on a false syllogism. Crudely put, this is the idea that quantum mechanics is mysterious and difficult to understand, and consciousness is mysterious and difficult to understand, so therefore consciousness must involve quantum mechanics.

What positive grounds might there be to invoke quantum mechanics to understand consciousness? I do not think there is any reason whatsoever to see any connection. Empirically, electrons are not delocalised anywhere in our nervous systems—remember that all of the electrical conduction that takes place in nerves results from the motion of ions by Brownian motion, and not by the coherent transport of electrons. There is no structure in the nervous system which you could plausibly imagine might have significant quantum effects, not least because every structure is continuously being bombarded by Brownian motion, destroying any quantum coherence on very fast time-scales. Some people seem to feel that the unpredictable outcomes implied by consciousness—related to the philosophical problems associated with free will—are inconsistent with the determinism of classical mechanics and need the fundamentally probabilistic character of quantum mechanics. But there is no need to invoke quantum mechanics to understand the stochastic quality of consciousness, as the Brownian motion of the many signalling molecules that leads them to diffuse across spaces to their receptors has quite enough in-built randomness. What about the difficulty in localising the phenomenon of consciousness in one particular spot? But this quality is shared by many other collective self-organising phenomena that arise from the non-linear interactions of many individual components.

One driving force for trying to introduce quantum mechanics into a theory of consciousness has been the realisation that quantum mechanics can, in principle, be used as the basis for a type of computing that is fundamentally different to our current methods, and potentially very much more powerful. Our current digital logic is based on the idea of storing and manipulating information using, as the basic unit, a bit, which can take one of two states—on or off, yes or no, 1 or 0. The basic unit of information in quantum computing, a quantum bit or qubit, can exist in a state corresponding to 1 or 0, like a classical bit. But it can also exist in a superposition of these states, in a mysterious and fundamentally quantum condition in which it has the potential to be in either of these states with some certain probability. If one can carry out operations on qubits in which this quantum superposition is maintained, then effectively the operation is being carried out on all of the possible states at once. A quantum computer is intrinsically massively parallel, with even a very simple quantum computer being, in principle, equivalent to astronomical numbers of ordinary computers working simultaneously.

The snag is finding a system in which quantum computing can be executed. The problem comes because the quantum superposition of states which we are relying on only persists for as long as the computer can be kept isolated from the environment. One very promising way involves isolating atoms at very low temperatures in an optical trap; quantum dots are another possibility, but only if they are cooled down to a few degrees above absolute zero. Only at these very low temperatures can we avoid the random collisions that lead to what is called decoherence—the unscrambling of the quantum superposition and the irreversible loss of the associated information.

On the face of it, it is difficult to imagine a less likely environment for quantum computing than the brain. Far from being close to absolute zero, our brains are at a temperature at which random Brownian motion is a dominant factor. Rather than being isolated from the environment, any possible location for quantum computing will be continually interacting with neighbouring molecules of water and other molecules. The proponents of theories of consciousness and the brain that involve quantum computing have a lot to explain before their theories are going to start to look plausible.

8
Single-molecule electronics

I recently went to a seminar by a distinguished chemist about an aspect of his work on nano-materials. At the end of the seminar, in response to a question, he began musing about why biology never seemed to use electron transport—as we saw in the last chapter, biological electrical phenomena are dominated by the movement of ions. 'Of course, biology doesn't use electrons, for the same reason that you shouldn't use a toaster in the bath', I thought to myself, but I was too polite to say anything so sarcastic. Which was just as well, because I soon realised that there was a lot more to this question than the simple incompatibility of electricity and water.

Living things are made from organic molecules (that is, molecules based on carbon), and until fairly recently this fact by itself would have been thought to rule out any significant role for electron transport. But physics in the last few years has grown a new field—organic electronics. It is now clear that, not only can organic materials show interesting electronic properties, but that it may well be possible to make useful devices from them. They raise a tantalising possibility—that we will be able to use these conducting and semiconducting organic molecules to achieve what would be a spectacular success for nanotechnology—creating electronic circuits in which the basic component is a single molecule.

So, if it is possible for organic molecules to have electrically interesting properties, then why does nature not use them? The answer is that it does, but only for harnessing energy, not for processing information. The major role that nature has found for coherent electron transport is in the process of photosynthesis, in which the energy of sunlight is used to make life's fuel.

The green goo catastrophe

If we were looking for a place in the universe in which life might evolve, then we would look for a small, rocky planet, big enough to sustain an atmosphere, at the right distance from its star for plenty of water to be present in its liquid state. Imagine such a planet, in which, after the cataclysms that followed its formation had calmed down, a form of life had evolved. More than a billion years pass, and from its early, crude beginnings, thriving communities of life-forms of astonishing diversity develop. Many different kinds of organisms have evolved to fill a complex network of ecological niches; some of them exploit reactive chemicals available in the environment, others have learnt to harvest the energy of sunlight. Some live passively in huge colonies, others have the ability to move in search of food or to escape poisons.

Imagine now that one day, one of these life-form evolves a new way of making a living of unprecedented effectiveness. By combining the energy of sunlight with one of the most common materials in the environment, it can produce food in a substantially more efficient way than its predecessors. Because it can use as raw material one of the most abundant substances on the surface of the planet, it soon spreads at the expense of older life-forms, coating the surfaces of rocks, lakes, and seas with its green, slimy goo.

But, in addition to these advantages, the new form of life has a devastating weapon—the by-product of its highly-efficient metabolism is a toxic gas, which is deadly in its effects on the older rival forms of life. The green goo can tolerate this poison, and shortly learns to exploit it, but the older rivals are driven underground. In time, the green goo has remodelled the whole environment to suit its needs; the toxic gas forms a major component of the atmosphere of the planet and the older, original forms of life have been virtually wiped out, surviving on in only a few inconspicuous niches.

This 'green goo' catastrophe, so reminiscent of the 'grey goo' scenario that radical nanotechnologists worry about so much, is exactly what happened on our planet Earth around two billion years ago, when the ancestors of modern cyanobacteria developed the modern form of photosynthesis, in which sunlight is used to split the hydrogen from water, combining it with carbon dioxide to produce sugars and carbohydrates. The toxic by-product, is, of course, oxygen gas. In the course of a few hundred million years the concentration of oxygen in the atmosphere increased from less than 0.1% to something approaching its current value of 21%. The older forms of life, which cannot tolerate oxygen, continue on as the anaerobic microbes, confined to a few restricted niches where they can avoid the toxic gas. All other life—plants, fungi, and animals, including ourselves—one way or another are descended from the green goo.

The green goo catastrophe was a turning point in the evolution of life. Before the oxygen crisis, evolution had concentrated on developing the nanotechnology of life, culminating in what is perhaps the most sophisticated of all biological nanotechnologies—the mechanism of photosynthesis. The response of evolution to the colossal environmental challenge posed by the increase in concentration of oxygen in the atmosphere is perhaps surprising and counter-intuitive. One might think that the rapid and almost complete take-over of the planet by a single type of life—the cyanobacteria—might have led to a future of dull uniformity. Actually, the opposite happened. Working with the same basic nanotechnological tool-set that had been developed in the first phase of life's evolution, the emphasis switched to the growth of complexity. Under the influence of the powerful driving forces of symbiosis and co-operation, first of all much more complex single-celled organisms were developed, and then these single-celled organisms started to co-operate to make multi-celled plants and animals.

What makes photosynthesis so novel and powerful as a biological nano-technology is the fact that it adds a new element to the basic repertoire of principles used by biology. Photosynthesis exploits the basic tricks of soft nanotechnology—self-assembly, molecular recognition, and Brownian-motion-activated conformational change. But there is a new element—the organised transport of electrons. In photosynthesis, nature has taken the first steps to developing molecular electronics.

Dyes and photosynthesis

When we put a log of wood on the fire on a winter's night, we convert organic molecules, like cellulose, which consist of polymeric compounds of carbon, hydrogen, and oxygen, into carbon dioxide and water. This process releases energy, some of which we feel as the welcome warmth of a glowing fire. What is going on here is that we are undoing the process of photosynthesis. The growing tree uses the energy of sunlight to split water into gaseous oxygen and hydrogen in a reactive form; this reactive hydrogen is combined with water and carbon dioxide to create the molecules of sugar, which serve both as energy stores and as the basic components of structural molecules like cellulose. The biochemistry of these reactions is complex, but at the heart of the process is the use of light to split water into hydrogen and oxygen.

Like so many other important processes in biology, photosynthesis takes place in a lipid membrane. This membrane has a highly-convoluted shape, but the topology is simple enough. Inside the cell wall of a photosynthetic bacterium is the photosynthetic membrane, which forms an enclosure within an enclosure. In a multi-celled plant, photosynthesis takes place in specialised organelles called chloroplasts, which have a similar structure to photosynthetic bacteria. In fact, these chloroplasts are believed to be the descendants of photosynthetic bacteria which formed a symbiotic partnership with the cells that enclose them.

Embedded within this inner membrane are the photosynthetic reaction centres, at the heart of the process of converting light to chemical energy. At the core of the reaction centre are a pair of molecules of the dye molecule chlorophyll—the molecule that makes plants green. The energy of light is used to excite an electron to a state of higher energy. The lower-energy state that the electron occupied before being excited is now vacant, and one can think of this vacancy as being a 'hole'—a place waiting to be filled either by the original electron, or by a different one. We can think of the result of this input of energy into the dye molecule as being the production of a pair of entities—an excited electron and the corresponding hole that it leaves behind.

This is exactly what happens when light falls onto any dye molecule—an electron–hole pair is created. What normally follows, though, is that the high-energy electron immediately falls back into the vacancy it left behind, filling the hole. The energy that is released by this process is used to produce light, which is re-emitted. It is this process of absorption of light to make an electron–hole pair, followed by re-radiation of light at the same wavelength, that leads to the vibrant colours that characterise good dyes. Some other related processes can happen, too. Sometimes, before the electron can recombine with a hole, it loses some energy as heat. When this happens, the light that is re-emitted has a different wavelength to the light that was absorbed, a process known as fluorescence. To our eyes, fluorescence is especially spectacular when a dye absorbs light in the ultraviolet part of the spectrum and re-emits it as visible light. This sort of ultraviolet-fluorescent dye glows white when illuminated by the apparently dark radiation of an ultraviolet light. You see these lights sometimes on dance floors, where they make the dancers' white clothes glow brightly—detergent manufacturers incorporate ultraviolet-fluorescent dyes into their washing powders to give clothing that 'whiter than white' look. Light re-emission and fluorescence may be good for clothes, but these processes are exactly what a photosynthesising plant needs to avoid. A plant needs to harness the energy that has been absorbed from sunlight before it has a chance to radiate away again. To do this, the electron–hole pair needs to be split apart before the electron can fall back into the hole.

In the photosynthetic reaction centre, in addition to the special pair of dye molecules in which the electron–hole pair is created, there are another three dye molecules that are held in a protein framework through the thickness of the membrane. The role of these dye molecules is to spirit the high-energy electron away from the special pair before it can recombine. This is done in a series of steps at astonishing speed; the first step, taking the electron out of immediate danger of recombination, with the resulting loss of energy, takes only a few picoseconds (a picosecond is one-thousand-billionth of a second). The final step of the process is the transfer of the electron to another dye molecule, that, rather than being tightly bound to the photosynthetic complex, is free to move. This step is much slower, taking a fraction of a millisecond. But, because the electron has been taken out of range of its hole and out of danger of recombination, this step can be taken at leisure. The mobile, energised dye molecule that results is used to carry out a series of biochemical reactions that have two effects. Firstly, hydrogen ions are accumulated inside the photosynthetic membrane, and the energy that is stored in this way is converted into the chemical energy of the energetic molecule ATP by the molecular turbine complex ATP synthase. Secondly, another energetic molecule called NADPH is created, which together with ATP is used to power the reactions that produce sugar molecules from carbon dioxide and water.

The photosynthetic reaction centre is a remarkably effective way of harnessing the energy that is stored in the excitation of an electron by light and converting it into chemical energy. But photosynthesis uses another mechanism to maximise the conversion of light into energy; rather than relying on the direct excitation of electrons by light in the special pair of dyes at the heart of photosynthetic reaction centre, a complex system exists to collect light over a wider area, to exploit light with a variety of different wavelengths, and to funnel it toward the reaction centre. This mechanism uses what are known as light-harvesting complexes or antenna complexes.

A light-harvesting complex consists of many dye molecules—tens or hundreds of them—bound to proteins in precise ways. The dyes consist both of the familiar green chlorophyll molecules, together with another class of dyes called carotenoids. Carotene is the dye that makes carrots orange. Because carotene is somewhat more stable than chlorophyll, when the leaves on deciduous trees die at the end of the summer the chlorophyll decays first, and the colour of the leaves changes from green to the oranges, reds, and browns that makes the autumn season so spectacular in temperate countries. From the plant's point of view, the reason for having differently-coloured dye molecules in the light-harvesting complexes is to maximise the efficiency with which the sunlight is used; different coloured dyes absorb the different wavelength components of white sunlight and make sure the whole spectrum of colours is used.

When light is absorbed anywhere in the light-harvesting complex, it creates an electron–hole pair. The excitation energy of this electron–hole pair needs to be transferred from dye molecule to dye molecule until it reaches the special pair of dyes in the reaction centre, where the electron can be split from the hole and carried away to safety. Again, this energy transfer is a race against time. About one nanosecond (one-billionth of a second) after the electron–hole pair is formed in chlorophyll, it will recombine, the light will be re-emitted and the energy wasted. In fact, the time-scale for energy transfer is a few tenths of a picosecond, allowing many such transfers to take place before the electron and hole have a chance to recombine. The mechanism for this energy transfer is a quantum mechanical process related to the van der Waals force between uncharged molecules; the result is that one can imagine the electron–hole pair as an entity that is rapidly passed from molecule to molecule.

The key to the operation of both the light-harvesting complexes and the photosynthetic reaction centres is the efficient and very fast transfer of both energy and electrons. In both cases, what these transfers rely on is the very precise alignment of the dye molecules, which makes the electronic states of those molecules overlap and interact in rather specific ways. What is it that holds these dye molecules in place? It is the self-assembling properties of proteins. The dye molecules are associated with a group of proteins, and it is the folding of these proteins into specific shapes that brings the different dye molecules that make up the light-harvesting complexes and reaction centres together at precisely the right separations and mutual orientations for these fast transfers of energy and electrons to occur. If slight mutations are made to the proteins, so that the sequence of amino acids is slightly changed, then the correct alignments can be destroyed and photosynthesis stopped.

In photosynthesis, then, a molecular machine is made by the usual biological processes of self-assembly. But a new element is added—the transport of electrons and electronic excitations. What evolution has done is to make an electronic circuit by self-assembly. The possibility of emulating this feat is what drives the field of molecular electronics. But before we move on to this subject, let us consider a subject even more directly related to photosynthesis—how can we convert the energy of sunlight into a usable form?

Clean power for all—non-conventional photovoltaics

To those of us who were children in the 1970s, the idea of an energy crisis is deeply ingrained. At the beginning of the decade, I remember the rolling power cuts that followed a devastating strike by coal miners; to a child, bathing in candle-light was thrilling, but the energy shortages paralysed the economy and brought down a government. Later that decade, the oil shocks caused by revolution in Iran led to widespread economic dislocation, and reminded us all that the oil that our lifestyles depended on was a finite resource that would, before very long, be exhausted. Sheik Yamani, the Saudi Arabian Oil Minister at the time, was a figure who seemed to wield more power than the rulers of many rich countries. But he made a comment that was surprising to many who were predicting a new age of energy scarcity: 'The Stone Age didn't end because we ran out of stones, and the oil age won't end because we run out of oil.'

Taking the long view, adding up the global balance sheet of the energy the Earth receives from the sun and the energy we use, we are a very long way from a real energy crisis. Most of our energy ultimately comes from the sun (nuclear power being the exception); fossil fuels like coal and oil represent solar energy that was stored by plants many millions of years ago and deposited underground, while wind power, hydroelectricity, and the power from solar cells exploit the solar energy that is arriving right now. Currently, our energy sources are dominated by non-renewable fossil fuels. But this is an issue of economics, technological capability, and convenience rather than a fundamental inability of the sun to supply us with enough renewable energy.

This point is emphasised by the figures. The world's total energy production was estimated to be 428 exajoules in 1998 (one exajoule is 10^{18} joules). Of this production, around 70% was in the form of fossil fuels and another 6% in nuclear. Direct solar power accounted for a tiny 0.009%; indirect renewable sources are dominated by hydroelectricity, accounting for 6.6% of energy production, and wood fuel at 6.4% (though this may not be strictly renewable as forests are being burnt more quickly than they regrow). But the small contribution of renewable energy to the world's energy budget is not because the sun does not shine enough. In fact, nearly three million exajoules of solar energy reaches the Earth each year, more than 7000 times more than our current energy needs. We use fossil fuels because it is convenient and cheap to do so, not because the renewable energy resources are not potentially available in sufficient quantity.

Plants themselves are fairly efficient converters of solar energy; they convert about 1000 exajoules a year. Note that this figure itself is significantly bigger than the total human energy budget, but it is still a lot less than the total amount of solar radiation falling on the Earth's surface. Of course, a great deal of radiation falls onto oceans, deserts, or icecaps, but even of the energy that does fall onto plants, quite a lot is wasted. The overall efficiency of the photosynthetic process is around 5%.

How well can we do in emulating photosynthesis by converting the energy of light from the sun into usable, stored forms of energy? Solar cells, or photovoltaic devices, convert light directly into electricity, while we can imagine other solar-conversion devices that would produce stored chemical energy, as plants do. One particularly attractive possibility would be to emulate photosynthesis in the sense of using light to split water into hydrogen and oxygen, but to extract the hydrogen in elemental form, to use as fuel. Conventional solar cells work on rather different principles to photosynthesis. They are already quite efficient, but at the moment they are too expensive to be competitive with fossil fuels for generating electricity. Some novel concepts for solar cells are closer in their mode of operation to photosynthesis, and offer the promise of being much cheaper to make than the conventional sort. At the moment, however, they are not efficient enough and also have rather limited lifetimes. If these problems could be solved then they could potentially make a big difference to our energy economy.

In a conventional solar cell, the absorption of light leads to the production of an energetic electron and a corresponding hole. In this, the operation of the solar cell is similar to the first stage of photosynthesis, where an electron–hole pair is also created. But in photosynthesis we have seen that the creation of the electron–hole pair is localised in a dye molecule. The electron and hole are subsequently split up before they can recombine by a process in which the electron is transferred, on a very fast time-scale, to other dye molecules. But this feature of photosynthesis, in which one part of the device is dedicated to the production of electron–hole pairs, and another quite different part of the device is dedicated to the transport of the electrons after they have been produced, is very different to what happens in a conventional solar cell. These cells are made from two layers of semiconductor (both layers are usually silicon, with one layer being doped with holes and the other with electrons) sandwiched between two electrodes. Electron–hole pairs can be formed anywhere in the semiconductor, and the electron is transported to the electrode through the same semiconductor that the original excitation takes place in. There is no separation of the two necessary operations of a solar cell, namely the production of the electron–hole pair and the transport of the electrons. What separates the electron–hole pair in a solar cell is the fact that, within the device, there is an electric field associated with the junction between the two layers of semiconductor.

Micheal Grätzel, of the EPFL in Switzerland, invented an entirely new type of solar cell which is similar in concept to photosynthesis in that the functions of generating an electron–hole pair and conducting the electron away are separated. The basis of the Grätzel cell is a layer of nanoparticles of titanium dioxide (see figure 8.1). Titanium dioxide is a cheap and abundant material that is widely used as a pigment in white paint. In paint, the size of the crystals of the pigment

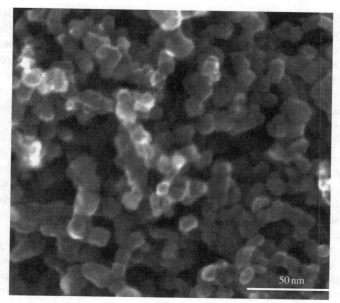

50 nm

Fig. 8.1 A porous film of nanocrystalline titanium dioxide, as used in the Grätzel solar cell. From Michael Grätzel. Photoelectrochemical cells. *Nature* **414** (2001) 338–44, with permission of the copyright holder, Nature Publishing Group.

is around a micron or so, but in the Grätzel cell the individual particles are much smaller, between 10 and 80 nm. The surface of the nanoparticles is coated by a single molecular layer of a dye. When sunlight is absorbed by the dye molecules, electron–hole pairs are formed. Just as we saw in photosynthesis, what needs to happen to stop the electron–hole pair from recombining, and thus re-emitting the light without capturing its energy, is that this pair needs to be separated. In the Grätzel cell, this is the role of the titanium dioxide; the electron is immediately transferred to the nanoparticles, from which it is transferred to the electrode. Meanwhile, lower-energy electrons are injected from the other electrode, from which they travel through a conducting material (an electrolyte) to the dye molecules on the surface of the nanoparticles.

The Grätzel cell is very different in the detail of how it works to photosynthesis, but they do share one common feature that separates them in principle from normal solar cells. The two operations, of electron–hole pair formation and electron conduction, are carried out by different materials. The fact that the titanium dioxide in Grätzel cells is so finely divided is important for two reasons. Firstly, since the conversion of the light energy takes place only on the surface of the titanium dioxide particles, one needs very finely-divided particles to create enough surface area to absorb a significant fraction of the sunlight. The other essential feature is that there is a continuous conducting pathway from every dye molecule to the electrode at which the electrons are conducted.

Grätzel cells are not the only novel approach to making solar cells that is currently being investigated. We will discuss later in the chapter solar cells made from conducting polymers, and a number of other configurations are being looked at as well. Do these have any realistic prospect of decisively shifting the economics of power generation in favour of photovoltaics?

The three key factors that are decisive in controlling whether solar cells are economically viable are efficiency, lifetime, and cost. The most efficient solar cells currently available are made from crystalline silicon; they can convert nearly one-quarter of the sunlight incident on them into electricity, but they are expensive. Cells made from amorphous or polycrystalline silicon are cheaper, but they are less efficient—the efficiencies are 13% for amorphous and 18% for polycrystalline silicon. These efficiencies are impressive, but it is the up-front cost of these solar cells (both in terms of money and in terms of the energy that has to be expended in making them) that stops electricity made from them being competitive with that made from fossil fuels.

The non-conventional cells are less efficient than silicon solar cells—for Grätzel cells, efficiencies are around 10%, while for cells made from semi-conducting polymers efficiencies approach 5%. Lifetimes can be a problem, too, particularly for polymer solar cells. But it is the prospect of making these cells at very low cost that drives the hopes of people researching them. Making silicon is intrinsically expensive and energy intensive, but the new kinds of cells are made by processes that are very easy to scale up to very large areas. The plastic foil that is used to make a crisp packet is not that different in principle from the sort of structure that will need to be used for a non-conventional solar cell; it is a base layer made up of a robust polymer, with a number of very thin layers of other polymers, pigments, and metals applied on top. While the largest silicon wafers that are currently made are tens of centimetres in diameter, production of coated polymer films is measured in hectares.

Both conventional solar cells and Grätzel cells convert light into electricity. Electricity is a useful and flexible medium of energy, but sometimes it is convenient to have our energy in a more portable form. Our transport system relies on stored chemical energy in the form of petrol/gasoline, so it might be valuable to be able to convert the energy of light into chemical energy, using it to make fuel. Again, this is closer to what photosynthesis does, in which the light is converted into such fuel molecules as ATP and energy storage molecules likes sugars and starches. Since the overall effect of photosynthesis is to split water into hydrogen and oxygen, it is tempting to look for a way of producing hydrogen directly from water and sunlight. Many people envisage a 'hydrogen economy', in which cars and lorries are propelled either by burning hydrogen directly or by using it in fuel cells to generate electricity. If the hydrogen was produced from solar energy, then such a hydrogen economy would be entirely renewable and would generate neither pollution nor carbon dioxide.

A simple, but effective, way of making hydrogen from sunlight would be simply to take the electricity generated from a conventional solar cell and use it to power the electrolysis of water. This works with something like 7% overall efficiency. Direct conversion schemes using nanocrystalline materials still cannot match this, but once again they offer the promise of being able to make large areas cheaply. One such scheme uses a modified Grätzel cell. Light is absorbed by nanocrystalline tungsten trioxide, an electron–hole pair is created, and the hole is filled by an electron taken from a water molecule. This splits the water into oxygen gas and positive hydrogen ions. Light of a different wavelength is absorbed by dye molecules adsorbed on the surface of nanocrystalline titanium dioxide, and the high-energy electrons can combine with the hydrogen ions to produce free hydrogen gas molecules.

Again, this mimics some features of photosynthesis at a conceptual level, but the details are very different. Can we make a more faithful copy of the apparatus of photosynthesis, featuring the light-harvesting complexes and the fast charge separation that make the process work so well? A group in Arizona State University have succeeded in doing just this; in their system, a synthetic light-harvesting complex and a synthetic photosynthetic reaction centre can be mounted in a phospholipids membrane by self-assembly (see figure 8.2). The reaction complex pumps hydrogen ions across the membrane, and the chemical energy that is stored in the gradient of hydrogen ions can be used to convert ADP to ATP using the natural turbine molecule ATP synthase.

This close copy of photosynthesis is not likely to be competitive with the more distant analogues, like Grätzel cells, when it comes to converting energy on a large scale for our power needs. But, if we do build nanoscale soft machines, then they will need a fuel supply, and this kind of blend of the natural and the synthetic could well meet that need.

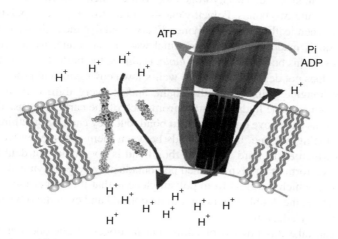

Fig. 8.2 An artificial photosynthetic membrane. A light-driven proton pump pumps hydrogen ions into the interior of a lipid vesicle; the resulting hydrogen ion gradient is used by the membrane-bound ATP synthase to convert ADP to ATP. Reproduced with permission from Devens Gust, Thomas A. Moore, and Ana L. Moore. Mimicking photosynthetic solar energy transduction. *Accounts of Chemical Research* **34** (2000) 40–8. Copyright 2000 American Chemical Society.

Organic metals and plastic semiconductors

It is difficult to imagine two types of material that seem so different as plastics and metals. Metals are shiny; plastics are clear or rather dully translucent. Metals are hard; plastics are tough. While the defining characteristic of metals is that they are good conductors of electricity, plastics are very good insulators. So the idea that a plastic could conduct electricity, perhaps as well as a metal, for a long time seemed quite ridiculous.

The discovery that changed this view was made in 1977. Hideki Shirakawa, in Tokyo, had discovered a way of making a rather uncommon polymer, polyacetylene, in the form of a film whose electrical properties could be conveniently measured. As made, the conductivity was not that high, but, on a visit to Alan MacDiarmid's laboratory at the University of Pennsylvania, Shirakawa and MacDiarmid experimented with exposing the polymer to iodine vapour. MacDiarmid's collaborator, Alan Heeger, found that this massively increased the conductivity of the polyacetylene. With a bit more refinement of the process, conductivities within an order of magnitude of silver and copper were obtained. For this discovery, Shirakawa, MacDiarmid, and Heeger won the Nobel prize for chemistry in 2000.

To understand how a carbon-based polymer can conduct electricity, we have to remember some elementary chemistry. The characteristic of a carbon atom (and the reason, that it is good at making polymers) is that it can make four strong chemical bonds with its neighbours. The simplest polymer—polyethylene—consists of a chain of carbon atoms, each linked to its two neighbours by a single bond. Two hydrogen atoms are attached to each carbon atom by its remaining two bonds. Polyethylene—familiar to everyone as polythene—is a tough, chemically-resistant polymer. Every bond that the carbon atom has at its disposal is made in a close-to-optimal way; its structure, as described by Primo Levi in his book *The periodic table*, is almost too perfect. The material is inert and does not degrade, leading to well-known environmental problems.

Now imagine going along the chain, and at every third carbon–carbon bond we remove one hydrogen atom from each of the carbon atoms. The two carbon atoms each have one unsatisfied bond, but they can do something about this by making not one, but two bonds between themselves. Now we have a chain in which every third carbon–carbon bond is not single, but double. This material is a form of a polymer called polybutadiene, a type of synthetic rubber. One way in which it differs from polyethylene is that it is rather more chemically reactive; the double bond is not quite optimal and even the oxygen in the air will slowly attack it.

We can take this process of taking out hydrogen atoms one step further. Imagine a polymer in which every carbon atom has only one hydrogen atom attached to it, so that every carbon atom has an unsatisfied bond. Now every other carbon–carbon bond will be a double bond. This polymer, called polyacetylene, in which the main chain is made from carbon atoms linked alternately by single and double bonds, is the material which Shirakawa discovered, and it is the first polymer to be discovered which could conduct electricity.

What does the alternation of single and double bonds along a polymer chain have to do with the way it conducts electricity? The explanation of chemical bonding lies in the way electrons are shared between atoms. In a single carbon–carbon bond, an electron state is created that is localised between the two atoms, and one electron from each of the two atoms is to be found in this state. In an isolated, double carbon–carbon bond, two joint electron states are formed, each of which houses one electron from each atom. But in a system in which single bonds and double bonds alternate, something special happens. This is called conjugation.

Conjugation has been known about for much longer than polyacetylene. The textbook example is benzene, in which six carbon atoms are joined together in a ring, each with a single hydrogen atom attached. According to the simple laws of bonding, we expect there to be an alternation of single and double bonds around the ring. But there is a problem with this—the molecule is completely symmetric, so why should one carbon–carbon bond be different to the next one? Experiments reveal conclusively that all of the bonds are exactly the same. The classical interpretation of this situation is that each bond has somehow a mixed character, partly single bond and partly double bond, a situation known to chemists as resonance. But the more sophisticated understanding of chemical bonding that arose when quantum mechanics became understood revealed that the situation was slightly more subtle than this. From each carbon atom, one electron is involved in a straightforward, localised bond; pairs of electrons are found in states that are localised between the two relevant carbon atoms. But one electron from each carbon atom is taken up into a different type of state—a state in which the electron can be found anywhere within two doughnut-shaped volumes above and below the ring of carbon atoms. These electrons are said to be delocalised, which means just what it says. Rather than being restrained in their motion to a volume quite close to their parent atoms, they are free to move anywhere within the molecule. Metallic behaviour requires delocalisation; in a metal the electrons are free to go anywhere within a macroscopic-sized piece of material. In benzene, the electrons are free to move too, but only within the confines of their little ring-shaped molecule.

Polyacetylene is like an infinite, linear version of benzene—the simple rules of bonding tell us that along its chains single and double bonds alternate. Electrons are free to move anywhere along the chain, so at least in one dimension we have something that begins to look like it might be a metal.

Conjugation and delocalisation occur at any time when following the simple bonding rules for carbon would lead to an alternation of single and double bonds between carbon atoms. In a polyacetylene molecule, it seems that the delocalisation should be of infinite extent. More local confinement occurs in benzene and dye molecules. Another material in which electrons are delocalised on a large scale is graphite. As we have discussed before, this form of carbon is composed of infinite sheets of atoms linked together in hexagons. Each carbon atom is joined to three other atoms. According to elementary chemistry, one bond is left unsatisfied and the electron associated with this bond is delocalised, free to wander anywhere in the sheet. Graphite is, indeed, quite a good conductor of electricity, as long as the electricity is travelling in the plane of the sheets rather than crossing between them; and, as a carbon nanotube is nothing more than a rolled-up sheet of graphite, these, too, have delocalised electrons.

In a metal, the electrons involved in electrical conduction are to be found in delocalised states. In the curious world of quantum mechanics that describes the properties of electrons, these conduction electrons are better thought of as waves than as particles; rather than being found in some definite position in space, they somehow pervade the whole volume of metal in which they are found. But having delocalised electrons is a necessary, but not sufficient condition for a material to be a metallic conductor. There is another condition, that follows from another, entirely quantum mechanical property that electrons have.

In Chapter 4 we talked about the different possible wave-like states that an electron could occupy. Just as, if you shake a bowl of water, patterns of water-waves form on the surface, there are different possible patterns of electron waves in materials. It is the peculiarity of electrons that only one electron can occupy one state. This is known as the Pauli exclusion principle, after Wolfgang Pauli, a pioneer of quantum mechanics. This principle has a profound influence on the way electrons behave in solids.

Imagine a block of metal sitting quietly on the table. If we had astonishing microscopic vision, we would see vast arrays of atoms arranged in crystalline order. The atoms would be a little blurred, as they would be vibrating a little bit, but the picture is essentially static. Now, if we look more closely to see what the electrons are doing, then we see a quite different picture. The electrons are delocalised—the position of each electron smeared out throughout the block of metal. If we were able to measure the speed of each electron, then we would find that they are moving fast—very fast. We associate the movement of electrons with the flow of electric currents—but we do not expect an isolated piece of metal to spontaneously develop an electric current. What is happening is that, although individual electrons are moving very fast, for every electron moving at a certain speed in one direction, another electron is moving in the opposite direction at exactly the same speed. All of the electric currents resulting from the fast movement of the electrons are exactly cancelled out.

Now, if we connect a battery up to the piece of metal, then we know that a current will flow. What happens is that some of the electrons that, before we close the switch, would be travelling from the positive terminal to the negative terminal, change their states so that they go in the other direction. Now there is a slight imbalance in the number of electrons going from left to right compared to those going from right to left, and a current flows. The crucial point that allows this to happen is that wave-like states are available for electrons to move into without too much input of energy. In a metal, there is a continuous series of wave-like states of gradually increasing energy, so that, when the rather small electric field that results from connecting up a battery is applied, a few electrons can change states, with the result that a net current flows.

If we had, not a block of metal, but a piece of a semiconductor, like silicon, then it turns out that there are no available electron states at energies only slightly higher than the energy of the fastest occupied electron state. In order to find a new state, a significant amount of energy needs to be applied. Semiconductors do conduct electricity, but not nearly as well as metals. But they can be made to conduct more electricity by a process known as doping.

In doping, we add a small amount of some impurity to the material. This impurity should have one of two properties. It should either have one extra electron compared to the atoms of the pure semiconductor, or it should have one fewer. If it has extra electrons, then these electrons need to be accommodated in the vacant higher-energy states. At these higher energies, there are plenty of available states, so when a field is applied these conduction electrons can easily change states to yield a net flow of electrons. A current flows, and this doped semiconductor conducts electricity. If the impurity has one electron fewer than the semiconductor atoms, then one of the possible electron states will be unfilled—there will be a vacancy. Just like those puzzles in which you slide square tiles around in a frame, the fact that one space is unoccupied means that it is possible to move the tiles around. In fact, one can think of the vacancy itself as being the thing that moves; in a semiconductor doped in this way, one can think of charge as being carried by a positive 'hole'. In polyacetylene doped with iodine, as discovered by Shirakawa, MacDiarmid, and Heeger, an electron is drawn out from the polymer chain onto the iodine, leaving a conducting hole.

The discovery of polymers that could conduct electricity led to great excitement, and people were enthusiastically predicting that cheap, lightweight plastic conductors would soon replace copper and aluminium power cables. The reality, as so often in technology, turned out very differently. A number of serious problems stood in the way of making useful products from what was a fascinating piece of laboratory science. The original material, polyacetylene, was impossible to process. Chemists have a name for this sort of material—brick dust. You cannot melt it, and it will not dissolve in anything; for anything useful to come out of this discovery, new methods of making it that would allow one to produce a useful film or fibre had to be developed. The breakthrough came at Durham University, where Jim Feast developed what came to be known as 'Durham polyacetylene'. We have already discussed this in Chapter 5; it is an example of what is known as a precursor route. The problem of processibility is overcome by preparing another polymer which can be processed, and which by a fairly simple process can be subsequently changed into polyacetylene.

Processibility was not the only problem; early conducting polymers were rather unstable, rapidly oxidising in the presence of air and losing their conducting properties. In order to get good conductivity, it was necessary to heavily 'dope' the polymer with other materials that are themselves highly reactive and toxic. Technology has progressed since then, and true metallic polymers are now commercial products, but they are used for rather small-niche applications, like anti-static coatings. The grand promise that polymers had for replacing traditional metallic conductors like copper and aluminium looks like it will never be fulfilled.

But, as always, what originally looks like a problem often turns out to be an opportunity. Undoped, or lightly doped, these materials are semiconductors— and, after all, silicon is more valuable than copper. The areas that conducting polymers are generating real excitement in at the moment exploit their semiconducting properties—making electronic devices like transistors, and optoelectronic devices like light-emitting diodes. If plastic electronics is with us now, then the potential for the future is that we use the semiconducting polymers one molecule at a time—making real the much-heralded field of molecular electronics.

Roll-up television screens and paint-on lasers

A solar cell collects light and converts it directly into electricity. The reverse trick is possible too—in a light-emitting diode, electricity is converted directly into light. This process is potentially very efficient. In a conventional incandescent light bulb, electricity is used to heat up a filament which then glows; most of the electricity is wasted as heat. But in a light-emitting diode working ideally, all of the energy you put in is converted to light. Light-emitting diodes made from inorganic semiconductors were originally used for displays on calculators, but developments in semiconductor nanotechnology mean that light-emitting diodes are beginning to be used for lighting purposes (for example, in traffic lights). This has the potential to reduce dramatically the amount of energy that the developed world uses.

The light-emitting diode was the first application of conjugated polymers that exploited their semiconducting character. Light-emitting diodes made from relatively small organic molecules that are semiconducting were discovered in the 1970s, but it was probably the discovery of light-emitting diodes made from a semiconducting polymer that had the biggest impact. The observation was made by Richard Friend's group at the Cavendish Laboratory in Cambridge in 1990. The first device consisted of a single, very thin layer of a conjugated polymer, poly(phenylene vinylene), sandwiched between two different metal electrodes. When a voltage was applied to the device a current started to flow, reflecting the transport of electrons and holes through the semiconducting polymer. But the student who was doing the experiment, Jeremy Burroughes, was observant enough to notice that the device emitted a faint glow.

What was happening in this early device was that electrons were being injected into the conjugated polymer on one side, and holes were being injected from the other. Most of the electrons and holes travelled straight through the material and carried their current to the other electrode. But, very rarely, an electron would meet a hole inside the polymer. Attracted to each other, the electron and hole would form a bound pair called an exciton, and in some of these excitons the high-energy electron would fall into the vacancy, releasing its energy in the form of light.

The efficiency of this early device was very small. Most of the electrons and holes simply never met and never had the chance to recombine and emit light. This very low efficiency meant that this prototype device would not be very useful; it had further problems in that the device was rather unstable, degrading both with use and with exposure to the oxygen in the atmosphere.

Since this first discovery, there has been a huge amount of development work that has brought these devices to the point of commercialisation. Synthetic chemists have developed new polymers which are more stable, and which emit light of different wavelengths. New ways of making more efficient devices have been devised. To make the devices more efficient, one has to arrange that every electron meets a hole, and every electron–hole pair recombines to emit light. The first requirement is to make sure that equal numbers of holes and electrons are injected in the first place. This can be done by choosing the electrodes carefully and by tailoring the structure of the interface between the electrode and the semiconducting polymer layer. To make sure that every electron finds its hole, you need to give them a venue to meet. The ideal meeting place is an interface between two different semiconducting polymers. A simple way of achieving this is to use a device made of two layers, but we can get a much greater area of interface if we use a mixture of two polymers, or indeed a copolymer, in which self-assembly during the process of making the film creates nanoscale domains of one polymer dispersed in the other.

With efficient polymer light-emitting diodes, which have useful lifetimes and which are available in three colours, we can think about making a full colour display for a computer or a television. What would be the advantages of this over other kinds of display? They come from the fact that polymers are plastics, with all their virtues (as well as some of their vices). They can be processed cheaply, they are flexible, and they can be made in large areas. The promise is of giant, roll-up television screens, at a price low enough to be almost disposable.

The grown-up brother of the light-emitting diode is the laser. In a light-emitting diode, the photons that make up the emitted light are produced at random, uncorrelated with each other, leaving the device in all directions. In a laser, by contrast, the photons leave as a coordinated pack, all going in exactly the same direction with exactly the same wavelength. Just as a disciplined army is much more effective than a rabble or a mob, the bright, coherent light of a laser has all sorts of uses that go far beyond what a light-emitting diode can do.

In the sixties and seventies, when people thought of lasers and what they could be used for, what came to mind would be something like a science fiction death ray—a beam of highly-directed energy that could be used for boring and cutting the toughest matter. Some high-powered lasers can be used for this sort of purpose, but the way in which lasers have quietly taken over our lives has been in transmitting information. There are at least three lasers in my house, not because I want to zap my neighbours, but because lasers form the mechanism for reading the information in CD players and DVD players. My telephone wires are still made of copper, but this is becoming rarer and rarer; large-scale information transmission is dominated by optical fibres, in which information is carried by pulses of laser light.

The lasers used in CD players and for telecommunications are semiconductor lasers, made not from silicon, but from materials like gallium arsenide, gallium nitride, and combinations thereof. Is it possible to make a laser from a semiconducting polymer? Two things are needed to make a laser—a suitable arrangement of electronic states, so that it is possible to maintain a large population of excited electrons (a population inversion), and a pair of mirrors, which traps the light in the cavity that they enclose, stimulating the emission of more and more light.

It has proved possible to generate a population inversion of electrons in a semiconducting polymer, and it is possible to enclose such a layer in a cavity between a pair of mirrors. There has not yet, however, been a definitive report of lasing in a semiconducting polymer driven by an electric current.

If a polymer-based laser is possible, then we would need to look for applications that exploit the special features of soft matter—in particular, self-assembly. In one of the most elegant forms of semiconductor laser—the vertical cavity surface-emitting laser—the mirrors are formed, not by thin layers of metal, but by a stack of alternating layers of two different semiconductors. If the thicknesses and refractive indices of the layers are chosen correctly for light of a certain wavelength, then interference effects between the light rays reflected from each interface in the stack make it behave like a perfect mirror. But in Chapter 5 we saw that block copolymer can self-assemble into complex structures, one of which would be just the sort of stack of layers that would behave as a perfect mirror. The question for the future, then, is whether we can make a laser exploiting both semiconducting polymers and the self-assembling properties of block copolymer. The ultimate would be a polymer layer that could be applied in one step, in which the molecules would automatically arrange themselves in the right way to make the complex structure of a laser—in effect, a paint-on laser.

A less speculative application for semiconducting polymers is as solar cells. As we observed at the beginning of this section, a light-emitting diode is essentially a solar cell running in reverse, and many of the engineering issues that have been solved in making efficient polymer light-emitting diodes are the same ones that would need to be addressed to make efficient solar cells. The potential advantages of polymer solar cells are the same as the advantages of Grätzel cells—the possibility of being able to make large areas cheaply. Time will tell whether the balance of the economics will shift enough to bring polymer solar cells into production.

Plastic logic

In the world of conventional, inorganic semiconductors, by far the largest volume of devices exist, not to convert light into electricity or vice versa, but to

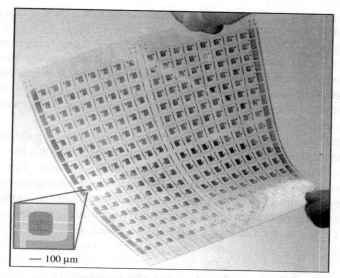

— 100 μm

Fig. 8.3 The backplane circuit for a e-ink application, consisting of transistors made from semiconducting polymers applied to gold electrodes which have been patterned by soft lithography. Reproduced with permission from John A. Rogers, Zhenan Bao, Kirk Baldwin, Ananth Dodabalapur, Brian Crone, V. R. Raju, Valerie Kuck, Howard Katz, Karl Amundson, Jay Ewing, and Paul Drzaic. Paper-like electronic displays: Large-area rubber-stamped plastic sheets of electronics and microencapsulated electrophoretic inks. *Proceedings of the National Academy of Sciences* **98** (2001) 4835–40. Copyright 2001 National Academy of Sciences, USA.

carry out logic operations and to store information in microprocessors and memory chips. The basic device that underlies both logic and memory in modern integrated circuits is the field-effect transistor. Semiconducting polymers can be used to make field-effect transistors, too. The basic principle is just the same as a field-effect transistor. A short length of the semiconducting material connects two electrodes. The conducting channel is separated from a third electrical contact—the gate—by a thin layer of insulator. When a voltage is applied to the insulator, extra charges are induced in the conducting layer of the semiconducting polymer. The more charge carriers that are present, the better it conducts—in this way, the gate voltage controls the flow of current through the channel like a tap opening and closing the flow of water.

The comparison in performance between silicon and a semiconducting polymer is still very much in silicon's favour. The perfection of the crystal lattice of the silicon that is used for micro-electronics means that charge carriers—electrons and holes—rush through the material with very little hindrance. A polymer, though, is a tangle of chains, which is inevitably undisciplined and disordered in its arrangement. Charge carriers take time to find their way through the mess (or, in slightly more technical language, they are frequently scattered by defects). Careful processing can improve things, by exploiting the tendency of these fairly rigid molecules to line up. But one is never going to be able to build anything that competes with a modern microprocessor on speed and performance. What might be competitive is the fact that a thin layer of polymers can be laid down from solution, at room temperature, and in ambient conditions—in short, without the colossally expensive infrastructure of a modern semiconductor fabrication plant.

To get these advantages, it is not sufficient to simply use a polymer as the semiconductor—each of the materials used, each of the layers that are put down to make the device, must similarly be made of plastic. This can be achieved; highly-doped semiconducting polymers can reach high enough levels of conductivity to act as the electrodes and connectors, while normal, insulating polymers can work well as dielectrics. All-plastic logic is now entirely feasible, and it is likely to be commercialised soon.

The ups and downs of molecular electronics

My colleague, Tony Ryan, tells me that when he teaches the subject of polymer processing, it is most important for the students to understand that polymers have three advantages—cost, cost, and cost. Plastics are cheap to make and they are cheap to process; their reputation for being the material from which shoddy and throw-away products are made is not without some element of truth. Certainly, if we look at the way the commercialisation of semiconducting polymers is proceeding now, then it is following the historic pattern of pursuing the cost advantage at the price of performance. Polymer light-emitting diodes are still less efficient than compound inorganic semiconductors, organic solar cells are much less efficient than silicon ones, and polymer logic circuits are painfully slow compared to the microprocessor in a PC. Actually, semiconducting polymers are not at all cheap when purchased by the kilogram, but the quantities used in an actual device, with an active layer of no more than a few tens of nanometres in thickness, are miniscule. What drives the emerging semiconducting polymer industry is the promise of cheap processing. At a time when semiconductor fabrication plants cost billions of dollars to set up, the idea that one can print out circuits using simple, cheap technologies like screen printing or ink-jet printing is very attractive.

But electronics using polymer and organic semiconductors has another potential advantage that has many people very excited—it seems to offer a simple way to achieve an unparalleled miniaturisation of circuits in which the circuit element is a single molecule. A single conjugated polymer would conduct electricity along its length just like a little wire; so, if we could properly connect up the individual molecules into logic circuits, then we would have the potential to make computers on the tiniest of scales. This seductive vision has become known as molecular electronics.

The idea of using molecules as the components of computers has been around a long time—since the early 1970s. The field has had its share of ups and downs. In the early days it was possibly guilty of overstating how quickly significant results were likely to come. In 1992 the distinguished theoretical physicist, John Hopfield—who has been responsible for important insights into the way biological computing works—stated that 'the field suffers from an excess of imagination and a deficiency of accomplishment'. At the time the assessment, while brutal, was probably fair. But, by 2001, the area was chosen by *Science* magazine (the leading American journal for reporting the most important advances throughout science) as the 'Breakthrough of the Year'.

But this triumph soon started to look a little hollow. Astonishingly, some of the most impressive of the results that contributed to this year of break-through were shortly revealed to be frauds. One of the centres of the new advances was the venerable American laboratory, Bell Labs, which was the origin of a series of apparently brilliant papers between 1998 and 2002. Bell Labs was the corporate laboratory of the Bell Telephone Corporation, at which, among many other ground-breaking advances of the twentieth century, the transistor was invented, the cosmic microwave background discovered, and the computer operating system UNIX designed. Changes in corporate America have meant that the laboratory has shrunk from its glory days; the break-up of Bell's monopoly hold on US telecommunications and the splitting-off of the local telephone companies (the 'baby Bells') have meant that the utterly reli-able streams of revenues and profits that had allowed Bell to sustain a huge fundamental science research activity for so long have come to an end. Bell Labs, now part of a free-standing technology company Lucent, has had to become very much more commercially focused, as well as being much smaller than it was in its heyday. Nonetheless, most observers thought that this torrent of papers were touched by the Bell Labs magic.

At the centre of this blaze of activity in molecular electronics emerging from Bell was a young German scientist called Jan Hendrik Schön. In the four years between 1998 and 2002, he published more than 100 papers, many in the most prestigious and high-profile journals, such as *Nature* and *Science*. The range of achievements was wide; they included demonstrations of super-conductivity in polymers and fullerenes. He reported making field-effect transistors from simple organic monolayers, and, most spectacularly of all, a field-effect transistor in which the conduction channel was composed of a sin-gle molecule. As one scientist put it, 'these guys are going to Stockholm'—a Nobel prize seemed assured.

But in the first half of 2002 doubts began to grow. The results were so spectacular and so important that many laboratories were trying to reproduce them. But, looked at more closely, the papers were tantalisingly short of the sort of nitty-gritty experimental detail that would allow another scientist to duplicate the results. Nonetheless, the principles seemed clear enough—was there some voodoo magic in the way that Schön had made his samples, some particular combination of settings on the sputtering machine that he had stumbled on by chance, that no one else could duplicate? For whatever reason, none of the laboratories trying to reproduce the results had any success.

Things took a sinister turn when it was noticed that some of the graphs in different papers looked strikingly similar, even though the measurements that were being reported were on quite different devices. On close examination, even the pattern of tiny wiggles in the curves that is caused by noise in the instruments was the same in the different graphs, even though this pattern should be quite random and should be different in every measurement.

Shaken by these growing doubts, the management of the laboratory engaged a panel of distinguished physicists to look into the results. Their report was damning. The original raw data did not seem to exist, the devices were not available for examination, and the records of the experiments were inadequate or non-existent. The inescapable conclusion was that the devices did not exist and the experiments had never been done—Schön had simply made the results up.

What can we learn from this rather shameful sequence of events? It certainly raises some hard questions about the way science is done. One point that has left many people nervous is the role of the co-authors of the papers. Science is done in groups, which have a hierarchy; the actual work in the laboratory is done by relatively junior scientists, like Schön. Senior scientists will perhaps help with the data analysis. They will often do some work drafting the papers, on which the whole group is listed as co-authors. But they will pretty much always, if the work is exciting enough, swan around the world giving talks about the work at conferences in congenial locations. To what extent are they responsible if the work, for which they are receiving the accolades of their colleagues, turns out to be fraudulent? Obviously, since, like most human endeavours, science is ultimately based on trust between co-workers, there is not much that can be done against out-and-out deceit. But the scientific community does have a right to expect that a senior scientist will have been interested enough in the work of junior colleagues to have wanted to see the data in its raw, unprocessed state, to have had discussions about the way the apparatus works, and to have seen what the devices actually look like.

Another striking feature of the Schön affair is the way in which the scientific community really seemed to want to believe these results. I would certainly include myself in this category; I eagerly read these papers and enthusiastically told my students about them in lectures, as examples of some of the most exciting work in physics at the time. Part of this predisposition to believe came from the power of the Bell Labs brand name—anyone with any knowledge of the history of this, perhaps the most famous of all physics laboratories, would be conditioned to expect results of the highest importance to come from there. But the other part came from some inherent plausibility in what Schön said he was achieving. Results from many other groups showed that molecular electronics really was starting to produce results as well as hype. The Schön affair certainly has given the credibility of the field of molecular electronics a blow, but it should not detract from the very real achievements and even more seductive promise of the field.

Single molecules as electronic devices

There are plenty of real achievements in molecular electronics. There are, perhaps, four stages that scientists need to reach before devices based on molecular electronics hit the market. They have to find materials that have the right electronic properties. This requires both synthetic chemistry and the testing of the electronic properties of the materials. For many purposes it is enough to test a great many molecules all at once; this certainly suffices to find out if the chemists are on the right track in what they are synthesising. But there are ways in which a single molecule behaves that are not captured by the average behaviour of very many molecules; so in the second stage one does need to pick out a single molecule and try out its electronic properties by itself. In the third stage, you need to be able to wire up a molecule in such a way that it produces a useful device, like a transistor. Finally, you need to be able to integrate many such devices with their associated wires and other components to make the equivalent of an integrated circuit—a molecular computer.

Scientists have gone well beyond stages one and two—finding the right materials and checking out their properties both in large numbers and one at a time. Good progress is being made on stage three, with some working devices now starting to appear. But the crucial stage four—integrating the devices to make a working system—is still in its infancy.

We have already talked about the requirements for organic materials to conduct electricity—the need for conjugation, so we have electron states that extend right through the molecule, rather than the electrons being bound closely to one or two atoms. The best candidates are very closely related to the kinds of polymers being used for light-emitting diodes and field-effect transistors on a larger scale than a single molecule. The major new innovation that is required to make a molecule potentially useful for true single-molecule electronics is some way of wiring it up at the ends—some sort of molecular connector to serve the same purpose as the crocodile clips (or alligator clips in the New World) that the electronics hobbyist uses to attach short lengths of wire to diodes, resistors, and transistors. The most promising method seems to be a piece of chemistry that we have already encountered in Chapter 3—the strong bond that is formed by a thiol group (the combination of sulphur and hydrogen) with a gold surface. Gold is of course ideal as a material for forming wires and connectors; so making a length of conjugated organic molecule with a thiol group at each end offers a good starting-point for single-molecule electronics.

These conducting molecules with sticky ends make it easy to characterise thin layers of molecules. If we put a gold-coated surface into a solution of these molecules, then they will neatly assemble in a single molecular layer, stuck down on the gold by their sticky ends. The result is known in the trade as a self-assembled monolayer, or SAM. By taking this monomolecular layer, and coating it with another layer of gold, one has a simple way of characterising the electronic properties of the very large number of molecules that constitute the monolayer.

(a)

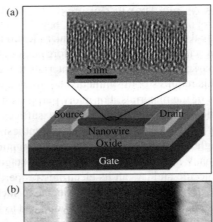

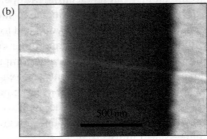

(b)

Fig. 8.4 A field effect tansistor made from a single semiconductor nanowire. (a) shows a schematic of the transistor; the inset shows a high resolution transmission electron micrograph of the nanowire. (b) shows a scanning electron micrograph of a the transistor; the nanowire can be seen crossing between the source and the drain electrodes. Reproduced with permission from Yi Cui, Zhaohui Zhong, Deli Wang, Wayne U. Wang, and Charles M. Lieber, *High Performance Silicon Nanowire Field Effect Transistors*, Nano Letters, **3**, 149–152, (2003). Copyright 2003 American Chemical Society.

The next step is to reduce the number of molecules being probed from many billions to a few tens or hundreds. The scanning tunnelling microscope is the ideal tool for this; it is a sharp point with an electrical connection whose position can be controlled with nanometre resolution. If we take a monolayer of conducting molecules deposited on a gold substrate, bring a scanning tunnelling microscope tip into gentle contact, and then complete the electrical circuit between the gold substrate and the tip, then we can start looking at, if not a single molecule, a handful of molecules. We can reduce the number of molecules that we are probing even further by making a monolayer that consists, not just of conducting molecules, but in which the conducting molecules are diluted by many non-conducting molecules.

But the most convincing demonstrations of measuring the movement of electrons through a single molecule use an ingenious method called a 'mechanically-controlled break junction'. The idea behind this is to make a weak link—a very thin gold wire, made by electron beam lithography—that connects two electrodes. The wire is free-standing, and is held a few nanometres above a rod. If the rod is bent, then the wire will break. In practise, a piezo-electric crystal is used to press the rod against a support, bending it by only fractions of a nanometre at a single step. When the wire has broken, but its ends are still less than 2 nm apart, then a solution containing conducting molecules with thiol end groups is dropped onto the bar. The thiol groups stick to the gold, and if you are lucky then you get one molecule that bridges the gap. This kind of break junction is very useful for physics studies, but of course it is difficult to see how you could make a commercial technology from it.

What is one looking for when one studies conduction in a single molecule—how does it differ from the behaviour of many molecules? One of the new features that arises follows from the fact that electric charge comes in units of an electron. If you put one extra electron into a macroscopic piece of metal wire, then it hardly makes any difference—it is like an extra drop of water in an ocean. But, if you inject an extra electron into a single molecule, that does not have that many electrons in it in the first place, then it is more like emptying a bucket of water into a basin that is already half full—it makes a significant difference to the state of the molecule, and means that it is that much harder to add a second electron. This feature is known as Coulomb blockade. Another possibility is that the action of passing a current qualitatively changes the shape of the molecule, and alters the way it conducts electricity next time. This is the basic action of a switch. You might be able to control whether the molecule passes a current when a low voltage is applied to it by applying a higher voltage, which would switch it from a conducting to a non-conducting state.

Fraser Stoddart and James Heath, at UCLA, have pioneered the use of a type of molecule with an intuitively appealing kind of switching behaviour. These molecules are called rotaxanes, and they consist quite literally of a ring threaded on a wire. The idea of this kind of molecule is that the switching event consists of moving the ring from one part of the molecule to another, making it an electrically-readable, single-molecule abacus. If we can find a way of individually writing to and reading from these molecules, then it is possible to imagine using them as a form of memory in which each bit of information is stored on a single molecule.

Wires and switches have only two terminals, but a transistor has three. There are the two terminals between which the current flows—the source and the drain—and a third terminal whose input controls this current flow—the gate. Conventional computer architectures rely on three-terminal devices, but they are very much more difficult to make than two-terminal devices. It is not too difficult to bring together two needles and press their points together; on a small scale what we have are two hemispheres, and two spheres will always meet at a point. But, if we need to bring a third needle point into contact with the points of the other two needles, then there is no alternative to very precise positioning. Most success so far with three-terminal devices has been obtained, not with organic semiconductors, but with inorganic semiconductor nanowires and carbon nanotube, though the first reports of single-molecule organic transistors are beginning to appear.

Integrating single-molecule electronics

No matter how ingenious the devices one can make using molecular electronics, unless they can be wired up to make a useful device, like a memory chip or a central processing unit, they will be as useless as a drawer full of old vacuum tubes. One very popular view is that the first applications of molecular electronics will occur in the form of hybrid devices. These are devices in which the framework of the circuit—the wires and the interconnects—are laid down using conventional planar semiconductor technology, and then the molecular circuit elements are deposited on this framework. This approach has the advantage that one can see how it might be possible to implement it, but the disadvantage is that it is not clear why this is going to be that much better than simply making the whole circuit out of silicon in the normal way. If one is going to have to use advanced lithographies such as electron beam lithography to fabricate feature sizes of a few nanometres in order to be compatible with molecular electronics, then it is difficult to see that using organic molecules rather than silicon as the semiconducting element is going to provide a massive advance.

A much more ambitious, but less well-developed, plan is to exploit self-assembly to get the circuits to make themselves. We have seen how organic molecules—particularly biological ones like DNA and proteins—spontaneously arrange themselves into complicated and quite specific three-dimensional structures. Can we design molecules with carefully-controlled interactions—precisely-located sticky patches and lock-and-key interactions—that will allow them to come together in exactly the right way to make a useful circuit? One would probably need to take a modular approach—assembling basic components like field-effect transistors as an assembly of a handful of molecules held together by weak physical forces in exactly the right configuration. This may sound far-fetched, but it is probably no more demanding than the precise way in which the dyes that make up the natural photosynthetic reaction centre are held precisely in place by the weak interactions between proteins and a lipid membrane. These modules could, then, perhaps, be wired up by interactions with gold nanoparticles.

Integration of molecular electronics by self-assembly is a demanding prospect, and many new ideas about circuit architecture will be required to make it work. These organically-grown circuits will be the antithesis of semiconductor integrated circuits, with their supreme reproducibility and precision—the vagaries of Brownian motion will ensure that every circuit is different, and ways of self-testing each circuit, even training it, will be essential. This vision of self-assembled molecular electronic circuits represents an approach to design that follows the spirit of biology—but the goal is something that evolution itself has not been able to achieve. That is certainly an ambitious target!

Meanwhile, in the related field of plastic semiconductors, academic and corporate pioneers are following the classic trajectory of a disruptive technology. The problem they face is the size, power, and history of success of the conventional semiconductor industry. It is all too easy for an academic with a good idea to compare the potential of a new information technology with the reality of computing now. But how do you keep up with the relentless juggernaut of Moore's law—the doubling of computing power every eighteen months? Conventional silicon micro-electronics is a moving target, and it is moving very fast. The only way to compete is not to compete. The approach that is being taken by the people commercialising plastic electronics is to chase niche markets that have only small value to the electronics mainstream. In this way, they can get some modest cash flows, and in the process they will be developing an infrastructure. The synthetic chemistry needed to make semiconducting organic materials will be being developed, experience in handling the materials on an industrial scale will mount up, and the fundamental science will move forward. It will be interesting to see if the very applications-focused world of plastics electronics and the visionaries of molecular electronics can combine to make this potentially revolutionary, radical nanotechnology really work.

9
Our nanotechnological future

Perhaps the most radical visions of the nanotechnology evangelists will not come to pass. Nonetheless, nanotechnology, in the form of machines structured on the nanoscale that do interesting and useful things, will certainly play an increasingly large part in our lives over the next half-century. How revolutionary the impact of these new technologies will be is difficult to say, and I am conscious that scientists have a very bad track record in prediction. Scientists almost always greatly overestimate how much can be done over a ten-year time period, largely because most scientists are quite unaware of the scale of effort and ingenuity required to take a viable idea from the laboratory and turn it into an engineering and economic success. But equally, scientists typically underestimate what can be done in fifty years. Working scientists are often cautious, incremental people, and their time horizons are as often determined by the typical duration of a grant from a funding body—three years—as by some long-term goal or vision.

But, even if we are fired by a grand vision of nanotechnology, we should not automatically be dismissive of the incremental. Science and technology does make progress by small steps. Sometimes the contrast between the grand visions of nanotechnology—robot nano-submarine navigating around our cells and repairing the damage—and the reality it delivers—say a better quality combined shampoo and conditioner—has a profoundly bathetic quality. But the experience gained in manipulating matter on the nanoscale in industrial production quantities is going to be invaluable. Similarly, there is no point being superior about the fact that lots of early applications of nanotechnology will be essentially toys (whether for children or grown-ups). But these apparently frivolous applications will provide the incentive and resources to push the technology further.

But we have seen that there is more than one design philosophy that could be used to develop nanotechnology far enough to meet at least some of the radical ambitions. Which of these philosophies is likely to deliver results?

Which way for nanotechnology?

I can identify at least four distinct approaches that one might hope to pursue in order to achieve some of the goals of radical nanotechnology.

The first approach basically develops the existing technologies that have helped push micro-electronics toward being nano-electronics, and micro-electro-mechanical systems toward nano-electro-mechanical systems. This top-down, hard nanotechnology uses developments of the planar technology of photolithography and etching to make devices on ever-smaller scales.

The advantages of this approach are obvious. A massive amount of existing technology and understanding is already in place. The investment, both in plants and in research and development, is huge, driven by the vast economic power of the electronics and computing industries. The obstacles to the further miniaturisation of the technology are already well understood and systematic efforts are in place to overcome them.

What disadvantages could this approach have? The first is that there are both physical and economic bounds on the ultimate lower limit in size that this technology can reach. The industry has shown extraordinary ingenuity in overcoming seemingly insurmountable barriers already, but maybe soon its luck will run out. Notwithstanding this, it is undeniable that this technology and the massive increase in computer power and information storage capacity that have resulted from it have already had, and will continue to have, huge and perhaps unforeseeable consequences on our society.

The second potential disadvantage is a more fundamental one. The only true example of a nanotechnology we have available to study is cell biology, and the major theme of this book is that biological nanotechnology works so well precisely because it embraces and exploits the peculiar features of the nanoworld. It exploits the combination of ubiquitous Brownian motion and strong surface forces by using self-assembly to fabricate, in parallel, huge numbers of precisely-configured nanoscale structures and machines. The combination of Brownian motion and biology's use of relatively soft and floppy structures together yields the principle of conformational change in response to changes in the environment as a basic mechanism for all kinds of machines, including molecular motors, gated ion channels in membranes, and cell-signalling events. The combination of small sizes and ubiquitous Brownian motion makes chemical diffusion a perfectly viable method for moving both material and information around the cell.

How could we follow biology's example, and work with the grain of the nanoworld? The most obvious method is simply to exploit the existing components that nature gives us. One way would be to combine a number of components that we have removed and isolated from their natural habitats, such as molecular motors, possibly incorporating them into artificial nanostructures. These artificial structures could be either soft, like artificial liposomes, or hard, made by silicon micro-engineering. Another approach would be to start with a whole, living organism, most likely a simple bacteria, and genetically engineer a stripped-down version that only contains the components that we are interested in.

One can think of this approach—often called bionanotechnology—as the 'Mad Max' or 'Scrap-heap challenge' approach to nano-engineering. We are stripping down or partially reassembling a very complex and only partially understood system to get something that works. An analogy might be what a technologically-backward group of people might do if they encountered a junk-yard of scrapped cars. Even without having the depth of understanding necessary to build a car from scratch, they might, after careful examination of the wrecks and by taking a few of them to pieces, work out how to combine the components in a way that produced something useful, even if it was only a wheelbarrow or a handcart.

The advantages of this approach are obvious. Nature has had both plenty of time and the use of the remarkable optimisation tool that is evolution, and has produced very powerful and efficient nano-machines. We now understand enough about biology to be able to separate out the cell's components and, to some extent, to run them outside the context of a living cell. This approach is quick and is probably the most likely way that we are going to get a full nanotechnology to work soon.

Given the attractiveness of this approach and its likelihood of producing quick results, what could the disadvantages be? I think one major disadvantage will be a reluctance by the public to accept such a technology. The history of the technology of genetic modification teaches us that, for better or worse, people tend to feel distinctly queasy about technologies that appear to rearrange or manipulate the basic elements of living things in quite such a fundamental way. A more concrete rationalisation of this distrust is the feeling that unpredictable consequences can follow from rearranging a complex system that is not fully understood. The recent history of gene therapy offers us a cautionary tale. To introduce genes into patients, it has been necessary to use what is essentially a piece of bionanotechnology—an adapted virus is used to introduce the new genetic material (that is, a stretch of DNA) into the cell. But fatalities have occurred due to unexpected interactions of the viral vector with the cell.

Can we retain some of the advantages of bionanotechnology—in particular, the design philosophy that exploits, rather than fights against, the special features of the nanoworld—while avoiding the potential disadvantages of using the components of living systems? One approach to doing this one might call *biomimetic nanotechnology*; that is, copying the principles of operation of biological nanotechnology, but executing them in synthetic materials. This seems to me to offer considerable advantages.

It is certainly less likely to produce unanticipated consequences than the full bionanotechnology route, and it may well prove easier to get public acceptance for. It is also easy to see some early potential successes using rather crude technology—for example, systems to keep molecules wrapped up until they are needed, and then to deliver them following some kind of trigger. In the longer term, the effort will be very helpful in our efforts to better understand the mechanisms that life uses at the nanoscale—there is no more effective way of understanding how something works than trying to build a copy.

On the other hand, the task of copying even life's simplest mechanisms is going to be formidably hard. We are lacking some of the crucial design tools—particularly evolution. To design molecules with a useful level of complexity—taking proteins as the bench-mark—we will need to find some way of incorporating evolution into our synthetic chemistry, whether that is by designing molecules to start with using computer-based evolution, or by designing direct evolutionary synthesis methods. For some of the most important technologies that cell biology uses—particularly in signalling and chemical information processing—we have not even begun to make synthetic analogues.

What then of the road to nanotechnology exploiting rigid diamondoid structures put together by positionally-sensitive mechanochemistry, as proposed by Drexler and his followers? I do not think that this approach fundamentally contradicts any physical laws, though I think that some of its proponents underestimate the problems that features of the nanoworld, like Brownian motion and strong surface forces, will pose for it. But I see many more disadvantages than advantages. Unlike the top-down route provided by planar technologies in silicon, there is no large base of experience and expertise to draw on, and no big economic pressures driving the research forward. Unlike the bionanotechnological and biomimetic approaches, it is working against the grain, rather than with the grain, of the special physics of the nanoworld. For these reasons, I anticipate that this approach to nanotechnology is least likely to deliver results.

What should we worry about?

Let us assume that nanotechnology in its radical form is possible and feasible—should we even want these developments to take place? Fifty years ago it might have been taken for granted that scientific progress was always positive in its effects, but it is certainly not now. Many people now wish that the scientific developments that led to nuclear power and nuclear weapons had not taken place, while more recent scientific developments, like the technologies of genetic modification of food crops, are deeply controversial. Now voices are beginning to be heard calling for caution in the development of nanotechnology; at the most extreme, there are demands for a complete moratorium on the development of the technology. There are many things about the development of nanotechnology that one could potentially worry about, but at the moment there seem to be two big concerns.

The first of these concerns is rather an immediate one, and it relates to the suggestion that finely-divided matter is intrinsically more toxic than the forms in which we normally encounter it. If the properties of matter are so dramatically affected by size, the arguments go, then does that not mean that matter which is harmless in bulk might be both more toxic and insidiously effective at getting into our bodies when it comes in the form of nanoscale particles?

The trigger for these worries is probably asbestos. This is a naturally-occurring nanoscale material which was extensively used in buildings and other applications in the second half of the twentieth century, but which caused substantial numbers of deaths by causing lung cancer and other diseases. One form of asbestos, chrysotile, consists of atomic sheets rolled up into tubes of a few nanometres in diameter. In another chemically-identical, but physically different form of the mineral, serpentine, the sheets are flat rather than rolled up; this material is completely harmless. The toxicity of chrysotile asbestos arises solely from its physical form.

Is this toxicity a general feature of nanoscaled materials? Attention is currently focused on carbon nanotube, which, like chrysotile, represent the rolled-up version of a sheet-forming mineral which itself is not toxic (graphite, in the case of carbon nanotubes). The analogy with asbestos is certainly close enough to suggest caution in the handling of nanotubes, but positive evidence of their toxicity remains to be found. On the other hand, we have sufficient experience of many other forms of nanoparticles to be sure that they present no special cause for concern. At the extreme, biological fluids like milk are full of nanoparticles, so we can be sure that there is no intrinsic reason to suppose that they are toxic. I think the sensible position must be that every new material has the potential to be toxic, whether it is structured on the nanoscale or any other scale, and we should make exhaustive studies to exclude this possibility before widely introducing such materials into the environment. We should certainly be aware of the fact that the properties of materials depend on their physical manifestation as well as their chemical content. But there is nothing to suggest that there is anything uniquely problematic about nanoscaled materials.

The other concern is the much more distant one, that is now irresistibly summed up as the 'grey goo' problem. Could we make, by accident or malevolent design, a plague of self-replicating nano-robots that spread across the biosphere, consuming its resources and rendering life, including ourselves, extinct? I hope that the reader of this book now realises what a difficult task making such a nano-robot would be. Nonetheless, the problem raises issues of principle which are worth confronting head-on. Let us first, however, consider some of the other potential impacts of nanotechnology which have not grabbed the headlines in the way that the 'grey goo' issue has, but which may still lead to far-reaching changes in society.

Some of these changes are happening already, without nanotechnology, but will be greatly accelerated if nanotechnology fulfils even part of its promise. Computing power will become more ubiquitous and much cheaper. Part of the promise of plastic electronics is that it will make computers so cheap that every artefact, even every piece of packaging, will have intelligence embedded within it, and the capability to communicate with other artefacts and to other computers. We are already beginning to see this happen with the spread of radio-frequency identification chips. Manufacturers and retailers love the idea of these devices, which will allow much tighter control of supply chains. But people concerned about civil liberties worry about the possibilities that these technologies hold for surveillance. Cheap, powerful computing combined with mass storage and automated image processing finally makes possible the totalitarian dream of being able to completely monitor a society, and the libertarian's nightmare of a total loss of privacy.

Enthusiasts are keen to promote applications of nanotechnology in medicine, and here expectations are high. Some people foresee nothing less than the complete conquest of disease and the end of ageing. It is going to take a while to reach this point, to say the least. On the one hand, our understanding of the detailed mechanisms of cell biology and their integration into life's complex interacting systems are still highly incomplete, while, on the other hand, we do not really have much idea yet of how we could make even one nanoscale robot capable of cell-level repair, much less the billions that would be required to make an impact on a pathological condition. But the foreseeable impacts of nanotechnology on medicine and medical science are big enough to merit serious consideration.

Once again, information is likely to be the currency that nano-medicine first brings us. Fast and cheap methods of sequencing an individual's genome, and instant tests for all kinds of biochemical imbalances—it is in diagnosis that we will see the first big impacts of nanotechnology in medicine. It is a matter of debate whether this will be entirely an unmixed blessing; too much information can sometimes be unwelcome. How does it help us to know that we are likely to suffer from some genetic ailment if the means to cure it are not available? All this kind of knowledge does for us is to make it impossible to get life insurance.

But even if we will not be able to mend our damaged cells one by one for a while, nanotechnology will certainly help us build replacements for damaged and diseased tissues and organs. Whereas now we are used to the idea of artificial legs and artificial hips, the development of tissue engineering, together with the development of nanoscale devices which might continuously monitor and correct our biochemistry, will lead to a much more intimate relationship between the natural and the artificial, the native and the improved. If some humans—the rich and powerful, most likely—are able to significantly enhance the capacities of their bodies by nanotechnology, then what frictions and conflicts will there be between the 'improved' minority and the unaltered masses?

Now we move away from the potential, concrete, social and economic impacts to those more intangible feelings that determine whether or not people are comfortable with a new technology. This goes beyond an accounting of the costs and benefits of the technology to a more intuitive feeling of acceptance and revulsion. Issues that concern the nature of life are particularly prone to lead to such a reaction—hence the gulf that has opened up between many scientists and many of the public about the rights and wrongs of genetic modification. These very profound issues about the proper relationship between man and nature are likely to become very urgent as bionanotechnology develops. Is it intrinsically wrong to take living organisms from nature to reassemble and reconstruct their most basic structures, possibly with additional synthetic components, in order to achieve our purposes? Is there a sense in which continuously replacing parts of humans by artefacts blurs the line between man and machine?

It is this concern about the proper relationship between man and nature that underlies the most far-reaching concern about nanotechnology—the suggestion that we will make self-replicating nano-robots that will escape our control and out-compete ordinary, natural life. Of course, this is a primal fear about any technology. The question is, is it realistic to worry about it?

We should be clear about what this proposition implies—the suggestion is that we can make an entirely synthetic form of life that is better adapted to the Earth's environment than life itself is. Can we out-engineer evolution? The practical answer is that this is certainly not possible now or in the next twenty years, and maybe not for a lot longer. We do not understand, at anything like the right level of detail, how life does work. We now have the parts list, but hardly any understanding of how everything fits together and operates as a complex system. Our appreciation of how nature engineers at the nanoscale will certainly grow fast, though, and attempts to mimic some of the functions of life will be very helpful in appreciating how biology operates.

But is it even possible in principle to develop a different form of life that works better than the one that exists now? To know this, we need to take a view on how perfectly adapted life is to its environment. How many times did life get started, and how many alternative schemes were tried and failed? Would any of these other schemes, which might possibly have been eliminated by chance or accident, have done better? Evolution is a very efficient way of finding the optimal solution to the problem of life. Does it always find the best possible solution? Maybe not, but I would be very surprised if we can do better.

Further reading

General references and Chapter 1

K. Eric Drexler. *Engines of creation*. Anchor, Garden City, NY, 1986 and 4th Estate, London, 1990.
(The original popular account of the vision of radical nanotechnology; still well worth reading.)

Richard P. Feynman. There's plenty of room at the bottom. Lecture at California Institute of Technology, 1959.
Available at http://www.zyvex.com/nanotech/feynman.html
(An influential and much-cited lecture anticipating the continuation of miniaturisation into the nano-domain.)

G. M. Whitesides. The once and future nanomachine; R. E. Smalley. Of chemistry, love and nanobots; K. E. Drexler. Machine-phase nanotechnology. *Scientific American*, 16 September 2001.
(This special issue on nanotechnology contained articles by Smalley and Whitesides that were highly critical of the radical vision of nanotechnology, together with an anticipatory rejoinder from Drexler.)

B. Joy. Why the future doesn't need us. *Wired*: 8.04, April 2000.
http://www.wired.com/wired/archive/8.04/joy.html

Will Self. *Dorian*. Viking, London, 2002.
(Origin of the quote in Chapter 1—this is the only mention of nanotechnology.)

Neal Stephenson. *The diamond age*. Doubleday, New York and Penguin, London, 1995.

Michael Crichton. *Prey*. Harper Collins, New York, 2002.
(Two very different novels whose plots revolve around nanotechnology.)

David S. Goodsell. *The machinery of life*. Springer-Verlag, Berlin, 1996.
(A brief but outstanding introduction to cell biology, whose illustrations give a vivid impression of the nano-domain.)

Bruce Alberts, Alexander Johnson, Julian Lewis, Martin Raff, Keith Roberts, and Peter Walter. *Molecular biology of the cell* (5th edn). Garland, New York, 2002.
(Biologists have good textbooks and this is one of the best. A clear, comprehensive, and beautifully illustrated account, at advanced undergraduate level, of the workings of biology at the nano-level.)

David S. Goodsell. *Bionanotechnology: lessons from nature*. Wiley, New York, 2004.

Philip Ball. Natural strategies for the molecular engineer. *Nanotechnology* **13** (2002) R15–R28.

Chapter 2

Felice Frankel and George M. Whitesides. *On the surface of things*. Chronicle Books, San Francisco, 1997.
(A beautiful scientific coffee table book with many striking images of the microworld and nanoworld.)

T. T. Perkins, D. E. Smith, and S. Chu. Direct observation of tube-like motion of a single polymer chain. *Science* **264** (1994) 819–22.

James K. Gimzewski and Christian Joachim. Nanoscale science of single molecules using local probes. *Science* **283** (1999) 1683–8.

Chapter 3

Neil Gershenfeld. *The physics of information technology*. Cambridge University Press, 2000.
(A technical, but succinct account of the fundamental physics underlying computing and information processing.)

Gary S. May and Simon M. Sze. *Fundamentals of semiconductor fabrication*. Wiley, New York, 2004.

Younan Xia and George M. Whitesides. Soft lithography. *Angewandte Chemie International Edn* **37** (1998) 550–75.

Harold G. Craighead. Nanoelectromechanical systems. *Science* **290** (2000) 1532–5.

Chapter 4

Steven Vogel. *Life's devices*. Princeton University Press, 1994.
(A general discussion of the way animals and plants are engineered to flourish in the physical world, with some very pertinent discussions of the effect of scale.)

Steven Vogel. *Life in moving fluids*. Princeton University Press, 1994.
(More detailed than *Life's devices*, this gives a quantitative but highly readable account of the fluid mechanics of swimming and flying.)

Philip Ball. *H_2O: a biography of water*. Weidenfeld & Nicolson, London, 1999.
(An excellent popular account of water's peculiarities.)

E. M. Purcell. Life at low Reynolds number. *American Journal of Physics* **45** (1977) 3–11.

Howard C. Berg. *Random walks in biology*. Princeton University Press, 1993.

Philip Ball. *Made to measure: new Materials for the 21st century*. Princeton University Press, 1997.

(An excellent introduction to modern materials science.)

Peter J. F. Harris. *Carbon nanotubes and related structures*. Cambridge University Press, 1999.

(Everything there was to know about nanotubes in 1999.)

Chapter 5

Philip Ball. *The self-made tapestry*. Oxford University Press, 1999.

(Discusses self-assembly as one example of spontaneous pattern formation in nature. An elegant and absorbing book.)

Richard A. L. Jones. *Soft condensed matter*. Oxford University Press, 2002.

(An undergraduate physics textbook with a lot to say about self-assembly.)

Ian M. Hamley. *Introduction to soft matter*. Wiley, Chichester, 2000.

(A good general introduction to self-assembling matter.)

Ken A. Dill. Polymer principles and protein folding. *Protein Science* **8** (1999) 1166–80.

(A review of what has come to be known as 'the new view of protein folding' from its main originator.)

Nadrian C. Seeman and Angela M. Belcher. Emulating biology: building nanostructures from the bottom up. *Proceedings of the National Academy of Sciences, USA* (2002) **99** 6451–5.

Chapter 6

Alexander Yu Grosberg and Alexei R. Khokhlov. *Giant molecules*. Academic Press, San Diego, CA, 1997.

(An outstanding popularisation by two renowned theorists.)

Allan S. Hoffman and twenty-eight others. Really smart bioconjugates of smart polymers and receptor proteins. *Journal of Biomedical Materials Research* **52** (2000) 577–86.

(How conformational change and molecular interactions can be combined for use in drug delivery and other areas.)

Ronald D. Vale and Ronald A. Milligan. The way things move: looking under the hood of molecular motor proteins. *Science* **288** (2000) 88–95.

L. Mahadevan and P. Matsudaira. Motility powered by supramolecular springs and ratchets. *Science* **288** (2000) 95–9.

K. Eric Drexler. *Nanosystems*. J. Wiley & sons, New York, 1992 (Drexler's detailed proposals for a radical nanotechnology based on nanoscale mechanical components made from diamond.)

Ricky K. Soong, George D. Bachand, Hercules P. Neves, Anatoli G. Olkhovets, Harold G. Craighead, and Carlo D. Montemagno. Powering an inorganic nanodevice with a biomolecular motor. *Science* **290** (2000) 1555–8.
(Powering an artificial structure made by lithography with a biological motor.)
Andrew J. Turberfield, J. C. Mitchell, B. Yurke, A. P. Mills, Jr, M. I. Blakey, and F. C. Simmel. DNA fuel for free-running nanomachines. *Physical Review Letters* **90** (2003) 118102.
Charles Singer, E. J. Holmyard *et al.* (ed.). *A history of technology. Vol. 4: The industrial revolution c.1750–c.1850, and Vol. 5: The late nineteenth century, c.1850–c.1900,* Clarendon Press, Oxford, 1958.
(Fascinating essays on the development of steam engines, machine tools, and many other areas of technology.)

Chapter 7

David J. Aidley. *The physiology of excitable cells (4th edn).* Cambridge University Press, 1998.
(A rather dense textbook about the nervous system.)
Christof Koch. *Biophysics of computation—information processing in single neurons.* Oxford University Press, New York, 1998.
A. G. Cairns-Smith. *Evolving the Mind.* Cambridge University Press, 1996.
(A popular account of the evolution of information processing in animals, from bacteria to consciousness. I like this book a lot, apart from the penultimate chapter.)
Leland H. Hartwell, John J. Hopfield, Stanislas Leibler, and Andrew W. Murray. From molecular to modular cell biology. *Nature* **402** (1999) C47–C52.
(A provocative essay discussing ways in which theoretical biology might proceed in order to understand the complex interactions that underlie phenomena like cell signalling.)

Chapter 8

David O. Hall and Krishna K. Rao. *Photosynthesis (6th edn).* Cambridge University Press, 1999.
Devens Gust, Thomas A. Moore, and Ana L. Moore. Mimicking photosynthetic solar energy transduction. *Accounts of Chemical Research* **43** (2001) 40–8.
Michael Grätzel. *Photoelectrochemical cells. Nature* **414** (2003) 338–44.
Richard H. Friend, R. W. Gymer, A. B. Holmes, J. H. Burroughes, R. N. Marks, C. Taliani, D. D. C. Bradley, D. A. Dos Santos, J. L. Bredas, M. Logdlund, and W. R. Salaneck. Electroluminescence in conjugated polymers. *Nature* **397** (1999) 121–8.

C. Joachim, J. K. Gimzewski, and A. Aviram. Electronics using hybrid-molecular and mono-molecular devices. *Nature* **408** (2000) 541.

James M. Tour. *Molecular electronics*. World Scientific, Singapore, 2003. (A technical monograph by one of the pioneers of the field. Heavily biased toward the work of his own group.)

Chapter 9

Anisa Mnyusiwalla, Abdallah S. Daar, and Peter A. Singer. 'Mind the gap': science and ethics in nanotechnology. *Nanotechnology* **14** (2003) R9–R13. (An interesting commentary on the state of the debate on the ethical and societal implications of nanotechnology.)

Stephen Wood, Richard A. L. Jones, and Alison Geldart. *The social and economic challenges of nanotechnology*. Economic and Social Research Council, Swindon, 2003.

Available from http://www.esrcsocietytoday.ac.uk/ESRCInfoCentre/Images/Nanotechnology_tcm6-1803.pdf.

Index